People and Power

People and Power
Electricity Sector Reforms and the Poor in Europe and Central Asia

Julian A. Lampietti
Sudeshna Ghosh Banerjee
Amelia Branczik

Cover design by Naylor Design, Washington, D.C. Photograph by Yuri Mechitov.

ISBN-10: 0-8213-6633-5
ISBN-13: 978-0-8213-6633-2
eISBN: 0-8213-6634-3
DOI: 10.1596/ 978-0-8213-6633-2

Library of Congress Cataloging-in-Publication Data
Lampietti, Julian A.
   People and Power: Electricity Sector Reforms and the Poor in Europe and Central Asia / Julian A. Lampietti, Sudeshna Ghosh Banerjee, Amelia Branczik.
   p. cm. – (Directions in Development)
Includes bibliographical references and index.
ISBN-13: 978-0-8213-6633-2
ISBN-10: 0-8213-6633-5
   1. Electric utilities—Europe.   2. Electric utilities—Asia Central.   3. Households—Energy consumption—Case studies.   I. Banerjee, Sudeshna Ghosh, 1973-   II. Branczik, Amelia 1978-   III. Title.   IV. Series: Directions in development (Washington, D.C.)

HD9685.E852L36 2006
333.793'2094—dc22   2006045424

Contents

Figures

Foreword

This important volume brings together a series of studies that were conducted to find out more about the distributional impacts of electricity sector reforms in the Europe and Central Asia (ECA) region between 1999 and 2004. At the time, there were serious concerns among policy makers and other stakeholders about the potential effects of the reforms on the poor, but precious little empirical evidence was available about exactly what these were and how best to mitigate them. These studies are an attempt to generate this information and identify more sophisticated mitigation strategies.

The studies are novel in the approach they take to analyzing utility reforms, and yield information that had previously been elusive and unavailable to policy makers. This provides policy makers with a more nuanced understanding of the effects of their reforms on the poor, and thus can improve their ability to mitigate adverse effects. Ultimately, this will improve the sustainability of reform and ensure that important macroeconomic objectives do not come at the expense of social development.

Beginning as a tool for understanding ex post the dynamics of reform, this approach was later used to produce a simulation of the effects of reform ex ante. This body of work thus provides fascinating insights into both the social effects of policies that have been implemented and the possible implications of putative reform efforts. In particular, the ability to forecast the effects of different policies is invaluable to the design of future reform (and the results and recommendations in these studies have indeed fed into subsequent reform design in several countries).

Though these studies focus primarily on electricity sector reforms, many of the themes run across the gamut of utility reforms. This makes the book an important contribution to the literature on the effects of infrastructure reform, particularly in the electricity and water sectors. The findings on the distributional impacts of cost recovery and coping mechanisms employed by households show us what happens at the household level when cost-recovery efforts are introduced. In addition, the book's revealing insights on the costs and benefits of different social mitigating strategies are an important contribution to the ongoing debate over subsidized utility provision versus direct transfers to the poor.

In contrast with other regions where reforms are aimed at increasing access to utility infrastructure, as a result of the Soviet legacy, countries in ECA have enjoyed almost universal access to electricity. These countries, therefore, face unique challenges in utility reforms that aim primarily at improving efficiency. By focusing on these challenges, this book fills an important gap in the literature on utility reform—and as countries in Latin America and elsewhere move closer to solving their access issues, the lessons of ECA will be increasingly relevant.

Breaking new ground at the time, the approach taken with these studies has been mainstreamed into Bank operations and is now routinely conducted to analyze the likely impact of policy reforms and determine effective mitigating strategies. Poverty and social impact analysis (PSIA) is now a vital input into the design of a broad range of reforms. It credibly informs policy makers and enables them to design social assistance mechanisms simultaneously with cost-recovery endeavors. This book highlights the potential of this approach, illustrates the kind of analysis that can be undertaken, demonstrates various ways of using and integrating quantitative and qualitative information, and offers invaluable guidelines for practitioners seeking to undertake such studies.

Following the Preface, which outlines the purpose of the book, Part 1 provides an introduction to the context of reform and the origin of the studies that form this book. Chapter 1 analyzes the background of crisis and reform in ECA, and the problems faced by policy makers as reform got underway. Chapter 2 gives a comprehensive overview of the methodology employed in these studies, setting it in the context of PSIA methodology.

Part 2 begins with an overview of reform patterns and changes in residential energy consumption in ECA during the 1990s, as energy sectors were transformed by crisis and reform. The four country case studies go on to reveal the factors that were at play in changing household behavior following the reforms. They offer detailed analysis on the effect of reform in their respective countries, and analyze the effectiveness of various mitigating strategies. Though they are united in examining the impact of reform on the poor, each case study highlights specific political economy conditions and sheds new light on questions of reform—the importance of understanding the effects of reform, problems associated with existing social benefit structures, and the importance of institutional factors in reducing nonpayment and improving cost recovery. The analysis of heating demand and the assessment of interventions in district heating in chapter 8 illustrate the importance of understanding heat as a major source of energy consumption in these cold climates, and household behavior patterns in designing infrastructure reform.

Finally, Part 3 brings together the findings in the case studies, offering a more in-depth analysis of some of the themes that have appeared in the preceding chapters. Chapter 10 concludes with an overview of the book's main findings, and offers broad guidelines on how to design effective reform, deal with exogenous factors, and mitigate the social effects. It also looks at lessons learned for analyzing reform, and offers guidelines for practitioners who are preparing to undertake similar analysis of infrastructure reform using PSIA.

This book is both a significant contribution to the literature on utility reform, assisting those who seek to understand the effects of these reforms, and an invaluable guide for those designing infrastructure reforms, in ECA and elsewhere.

Laura Tuck
Sector Director, Europe and Central Asia Region Environmentally and Socially Sustainable Unit (ECSSD)
January 2006

Acknowledgments

This book is based on six studies that were authored or edited by Julian A. Lampietti within the World Bank between 1999 and 2004. These included four country studies on Armenia, Georgia, Moldova, and Azerbaijan, and two regional studies, *Coping with the Cold: Heating Strategies for Eastern Europe and Central Asia's Urban Poor* (2002) and *Power's Promise: Electricity Reforms in Eastern Europe and Central Asia* (2004). The studies were edited and adapted for this book by Julian A. Lampietti, Sudeshna Ghosh Banerjee, and Amelia Branczik, who also wrote new material for chapters 1, 2, and 10. The book benefited greatly from editing by Bruce Ross Larson.

People and Power was generously sponsored by the Energy Sector Management Assistance Program (ESMAP). Within ESMAP, Douglas F. Barnes was invaluable in guiding and supervising our efforts from conception to completion, and we are extremely grateful for his support, and to Marjorie K. Araya for her guidance during the publications process. Valuable comments were provided by our peer reviewers: Robert Chase, Louise Cord, David Kennedy, and Johannes Linn. We are also very grateful for the comments and input of Lee Travers, Aline Coudouel, Laura Tuck, Peter Thomson, Nataliya Pushak, and the authors of the original studies: Julia Bucknall, Peter Dewees, Jane Ebinger, Irina Klytchnikova, Taras

Pushak, Gevorg Sargsyan, Sergei Shatalov, Katelijn Van den Berg, Anke S. Meyer, Anthony A. Kolb, Sumila Gulyani, Vahram Avenesyan, Ellen Hamilton, Hernan Gonzalez, Margaret Wilson, Sergo Vashakmadze, Nils Junge, Nora Dudwick, Karin Fock, Xun Wu, and Maria Shkaratan. The support of Lazlo Lovei, Alexander Marc, and Anis Dani was pivotal in the conception and execution of the original studies. Countless other colleagues within the World Bank, local consultants, and individuals in government ministries, energy regulatory bodies, utility companies, and community associations made instrumental contributions of research, data, and advice to the studies. Several of the studies also benefited from financial support provided by the Poverty Window of the Norwegian Trust Fund for Environmentally and Socially Sustainable Development and the Italian Consultant Trust Fund. Parts of the material in chapters 2 and 10 are based on a chapter written by Nils Junge and Julian A. Lampietti for *Poverty and Social Impact Analysis of Reforms: Lessons and Examples from Implementation* (Coudouel, Dani, and Paternostro 2006). The authors are also grateful for the supervision and guidance of Laura Tuck and management in ECSSD.

Abbreviations

ANRE	National Energy Regulatory Agency—Moldova
Btu	British thermal unit
CGE	Computable general equilibrium
CHP	Combined heat and power
CIS	Commonwealth of Independent States
CPI	Consumer Price Index
DALY	Disability Adjusted Life Years
EBRD	European Bank for Reconstruction and Development
ECA	Europe and Central Asia
ECSSD	Europe and Central Asia Region Environmentally and Socially Sustainable Unit (World Bank)
ESMAP	Energy Sector Management Assistance Program
EU	European Union
FSU	Former Soviet Union
GDP	Gross domestic product
GNERC	Georgian National Energy Regulatory Agency
GNI	Gross national income
GWEM	Georgian Wholesale Electricity Market

GWh	Gigawatt hour
HBS	Household Budget Survey
HH	Household
IFI	International Financial Institution
IMF	International Monetary Fund
KGOE	Kilograms of oil equivalent
KWh	Kilowatt hour
LPG	Liquefied petroleum gas
LSMS	Living Standards Measurement Study
NRED	Regional electric distribution companies, state-owned—Moldova
NTC	Nominative Targeted Compensation
PCE	Monthly Per Capita Expenditure
PPP	Purchasing power parity
PRS	Poverty Reduction Strategy
PSIA	Poverty and social impact analysis
RED	Regional electricity distribution
STC	Save the Children
UN	United Nations
UNDP	United Nations Development Programme
UNEP	United Nations Environment Programme
USAID	United States Agency for International Development
VAT	Value-added tax
WHAP	Winter Heat Assistance Program
WHO	World Health Organization

Note: All dollar amounts are U.S. dollars, unless otherwise noted.

Preface: Why Look at the Household Effects of Reform

The socialist legacy in Eastern Europe and Central Asia (ECA), where utility access had been extended to virtually all consumers at nominal cost, was an electricity sector leaching scarce fiscal resources from impoverished newly independent states, while seeing dramatic deterioration of its infrastructure. In the worst affected countries service was failing and electricity was unavailable for large parts of the day. The only option open in this situation was immediate implementation of a wide-reaching reform program.

The atmosphere of crisis that paved the way for reforms, and the urgency of reducing fiscal deficits and putting the energy sector back on its feet, precluded extensive consideration of the impact of reforms in advance. Reform was politically risky, but it was necessary—and it needed to begin immediately. The alternative, a collapse in utilities, was unthinkable. Those suffering most as a result of cost recovery, the poor, would be compensated, ideally with lump-sum transfers. When the momentum of reform began flagging, due to dissatisfaction with its perceived effects and mounting political pressures mobilizing against it, policy makers and the development community began to turn their efforts to understanding more about the concerns that mobilized opposition to reform. Although hostility to reform came also from those with vested interests in the status quo—ministers unwilling to lose their power bases, utility managers, and utility employee unions—consumer opposition to tariff increases lay behind some of the most virulent and vocal opposition, adding legitimacy to the antireform rhetoric of other constituencies.

The value of being able to separate perception and polemic from reality was obvious. What were the outcomes of reform? What were the effects on the poor? How could the design of reform, and mitigating strategies to soften negative impacts on the poor, be improved?

Although policy makers were searching for answers to these questions, no routine tool existed to analyze distributional impact—not only for privatization, but any policy reform. The World Bank had its poverty assessments, but they were not designed to answer these questions. They tended to be descriptive and their analyses of changes in poverty not policy specific. They were also of limited use in designing strategies to alleviate the effects of reform on the poor. Since they did not contain models for simulating responses to specific policies, it was almost impossible to measure empirically how different approaches to sequencing reform—such as increasing collections first, followed by raising tariffs—affected certain impacts groups. Without a tool or framework to examine distributional impacts on stakeholders, it was difficult to comprehend the aggregate picture, modify the design of reform, and devise a more effective social assistance strategy.

Against this backdrop, various studies were undertaken using different quantitative and qualitative techniques to answer some of these questions.[1] The studies that form the basis of this book, commissioned as part of the World Bank's analytic and advisory output contributing to this work, are based on the hypothesis that more careful attention to household preferences and behavior can smooth transition and reform of the power sector. These studies focus on quantifying the poverty and social impact of reforms. They identify what has worked and what has not in promoting both equity and efficiency, recognizing the importance of externalities, information asymmetries, rent-seeking behavior, and other attributes of imperfect markets. They look at households' coping mechanisms, the roles played by social assistance compensation, and consumer perceptions of reform. These studies were among the first examples of a new systematic analytic approach now widely used at the World Bank—the poverty and social impact analysis (PSIA, explained in depth in chapter 2), which aims to measure the distributional impact of major reforms on different groups in society, particularly the poor.

The contribution of these studies—and by extension this book—is threefold. By providing answers to the troubling questions raised by reform, the studies can help steer the future direction of reform, both in the respective countries and in the region, in a way that is responsive to the needs of reforming countries. Although a decade and a half has passed since the beginning of transition, much remains to be done in power sector reform in ECA (the 15 countries that emerged from the collapse of the Soviet Union and the 12 countries making the transition from socialism in Central and Eastern Europe).[2] Many countries still need to raise tariffs toward cost-recovery levels to make the power sector financially viable and encourage efficient resource consumption. Estimates indicate that residential electricity tariffs are below cost recovery in 14 of 19 ECA countries.[3] The sizable tariff increases needed are unlikely to be welfare neutral unless accompanied by substantial improvements in service quality or cushioned by appropriately designed income transfers. The lessons from these studies can inform the design of reform and the accompanying social policies—to maximize the welfare benefits and lessen the negative impact of tariff increases.

The studies were initially intended for policy makers in the ECA region, but the book informs the broader debate on the impact of power sector and utility reform, contributing to the literature on distributional impacts of infrastructure reform. While much has been written on this subject, the majority of studies to date look at Latin America; very few focus on ECA.[4] Yet ECA has important characteristics that sharpen our understanding of how different factors affect policy choices, particularly the starting point of universal access. For some countries, the challenge is to increase access while commercializing their utilities. But ECA's experience will be more relevant to economies that are moving toward fulfilling their access goals and will soon progress to service delivery challenges, such as those in Latin America. And given the similarities between the electricity and water sectors, there is substantial scope for learning lessons for water sector reforms.

These studies also illustrate the potential offered by PSIAs to understand the impact of reforms and improve their design. In recent years, the World Bank has placed more emphasis on understanding the poverty and social implications of reforms. Poverty reduction is now articulated as the main goal of the Millennium Development Goals, and development institutions emphasize a more country-owned rather than donor-driven approach to reform. Within the World Bank, a more "holistic" approach to development is institutionalized in Poverty Reduction Strategies (PRSs). Driven by the client country, these embody the World Bank's embrace of participatory development and are informed by PSIAs.[5] The studies in this book are some of the earliest examples of PSIAs, a framework that is now mainstreamed and embedded in the Bank's operational strategy.[6]

This book aims to provide insights into how household consumption and expenditure change in response to reform and what happens to payment levels, coping mechanisms used by households, and service quality improvements. It looks at the main strategies used by policy makers to mitigate the impact of reform and assesses the efficacy of these strategies in different settings. In the course of finding answers to these questions, it illustrates the key factors in the design of reforms that contribute to making them successful and examines how reform is affected by factors external to its design, including institutional and political economy factors.

Part 1 provides an introduction, looking at the promises and the problems of reform and the methodology used to assess them. The studies empirically measure the impact of reform by introducing several sources of data, most importantly the integrated use of data from household budget surveys and data on energy use and expenditure obtained from utility companies. By correlating the household and utility data for individual households, the studies generated more precise measures of how households responded to changes in energy price and supply and cross-checked the two sources.

The chapters in Part 2 are based on individual case studies. An introduction to energy consumption patterns in ECA in the past decade and a half in chapter 3 is followed by country case studies in chapters 4–7. Each case study sets the scene by looking at patterns of household energy consumption before focusing on one or more specific policy questions related to electricity sector reform. The analysis of the effects of electricity price increases on the poor in Armenia is the first case study (chapter 4). At the beginning of 1999, Armenia raised prices significantly and changed the structure of its tariff system from a tariff-based subsidy to a much higher uniform tariff accompanied by mitigating transfers to alleviate the impact on the poor. The study, conducted immediately after the reform, generated empirical evidence on how large the tariff increase was, who was most affected by reform and removal of subsidies, and how effective the transfers were in comparison with the subsidies they had replaced.

Chapter 5 looks at how households responded to tariff increases in Georgia. This study was conducted several years after reforms were put in place, giving a longer term perspective to the findings. By looking at how the utility attempted to increase payments, it sheds light on the role of institutions, government commitment, and the design of privatization in improving payments.

Chapter 6 considers Moldova, where a newly elected Communist government threatened to reverse one of the biggest privatizations in the region. One argument used by the opponents of reform was that it had disproportionately affected the poor and that privatization in particular had a negative effect. This led to a deeply acrimonious debate surrounding electricity sector reform and the sale of part of the distribution system to a foreign operator. This study provided ex post evidence that the accusations by opponents of privatization were groundless, thus answering an important question and showing how this work can improve public debates on reform.

Chapter 7 presents an ex ante study of reform in Azerbaijan, a country with markedly different circumstances. As an oil-exporting country, Azerbaijan was not faced with the same urgency to reform as the other three. The government's ambivalence about reform centered on the detrimental effects it could have on the poor and the possible political fallout of reform at a sensitive time in the presidential election cycle. In an attempt to inform policy discussions and lay out alternative scenarios for the government, the study looks at the welfare effects of different rates of increase in tariffs. It also estimates the level of compensation needed in each case to keep consumers as well off as before reforms. The study illustrates how the PSIAs can be used to design better reform strategies going forward.

Chapter 8 is a thematic case study, examining the most important aspect of energy consumption in ECA, heat. By once more examining trends in household consumption and demand, it suggests approaches for improving the traditional approach to designing investments in heating systems. It offers alternative recommendations on appropriate investments and policies to promote access to clean, affordable heat for the poor.

In reading the case studies, it is important to remember that they were conducted at a specific point in the timeline of reform and that the findings relate to the period for which data are analyzed, rather than for the reform period as a whole. It is their ability to give a picture of what is happening at a given point, rather than an evaluation of the reform program from inception to completion (most reform programs are not complete), that makes these studies valuable.

Part 3 synthesizes the lessons about the impact of power sector reform on the poor. Chapter 9 builds on the studies in Part 2 and on broader studies of household response to tariff increases across the region, and reflects on the implications for operational design of power sector reform in ECA and other regions. It reviews lessons on how to ensure that the poor are not disproportionately affected, with an analysis of the most effective mitigating strategies. Chapter 10 provides an overview of the book's main findings.

Notes

1. For a comprehensive bibliography of such studies see Foster, Tiongson, and Laderichi (2005), pp. 121–43.
2. This book uses the World Bank term "Europe and Central Asia" (ECA) to refer to the 27 former Soviet Union countries and the formerly socialist countries of Central, Eastern, and Southeastern Europe (the World Bank also includes Turkey in ECA, but this country is not included when referring to ECA in this book).
3. In percentage terms, the largest increases are needed in Central Asia (Azerbaijan, the Kyrgyz Republic, Tajikistan, and Uzbekistan). These figures are for 2003 and were calculated from World Bank ECA electricity data.
4. Foster, Tiongson, and Laderichi (2005), pp. 121–43.
5. This broader approach is known as the Comprehensive Development Framework, the emergence of which is widely associated with James D. Wolfensohn's tenure as World Bank president.
6. World Bank (2004d).

PART 1

Introduction and Methodology

CHAPTER 1

Power's Reforms—and the Problems

One of the most remarkable transformations of postcommunist Europe and Central Asia (ECA) was the mass reform and privatization of industry, infrastructure, and utilities that emerged from the economic collapse of the early 1990s. As with all reforms necessitated by crises of such magnitude, crises affecting the lives of so many in such tangible ways, the move to cost recovery as part of the fundamental restructuring of utility infrastructure was seen as either panacea or pariah of the new postcommunist economic and social order, depending on where people stood. For policy makers and economists, it was the only response available to a fiscal and economic crisis brought about by decades of manifestly unsustainable utility provision; the alternative was a collapse of the utilities. For consumers confronted with rising prices for energy and other utilities, it embodied the cataclysmic losses they were experiencing as part of the transformation of their social contract.

Utility reforms aimed at cost recovery and privatization have become one of the most divisive and politically charged economic issues of the past two decades. This book grew out of a desire for an empirical understanding of the effects of these reforms on the most vulnerable stakeholders: poor consumers of energy. A better knowledge of these effects, and how they come about, can provide lessons on how to improve the design of future reform to minimize welfare losses for the poor.

For most of the 20th century, utility infrastructure was generally the preserve of the state, in poor and wealthier countries around the globe. The natural monopoly characteristics of infrastructure networks, the large up-front investments required, the increasing returns to scale, the positive spillover effects of connecting all users to the network—all of these issues made infrastructure the natural responsibility of government. For political reasons, utility service delivery was often highly subsidized and available to consumers at below-cost prices. Supported by government largesse, state-owned utilities had few incentives to raise their own resources or improve the efficiency of their output. And in much of the world in the second half of the 20th century, state-managed infrastructure became synonymous with mismanagement, corruption, inefficiency, poor service, and huge fiscal transfers to cover operating losses. Donor-funded attempts to improve the record of state infrastructure in the developing world were repeatedly confounded by these structural characteristics.

In the 1980s, the role of the state was transformed as groundbreaking privatization schemes in the United Kingdom and Latin America heralded a drive away from government ownership of industry and infrastructure. As technological innovations—such as the ability to unbundle vertically integrated power utilities into separate generation, transmission, and distribution entities—made it feasible to introduce competition, operating these sectors as commercial ventures with private participation became a more realizable goal. By the 1990s, privatization of utility and physical infrastructure was gaining momentum and seen by many as a panacea for problems of infrastructure management.

At the same time, the World Bank and other international financial institutions (IFIs) became strong proponents of this approach in developing countries. Commercializing and privatizing infrastructure operations and introducing competition between different suppliers was seen as the most effective means to achieve the investment capital and efficiency improvements needed for sustainable utility sectors. Privatization had the added advantage of making reform politically feasible because it allowed governments, for many years pressured into providing cheap electricity to residential consumers and failing industries, to distance themselves from unpopular but necessary price increases.

By the early 1990s, government retreat from infrastructure had become a global phenomenon. Nowhere would this move be more dramatic than in the ECA region. Together these countries faced a common set of challenges in transitioning from socialist political and economic systems to market democracies, a process that is still ongoing. In the early 1990s, transition involved political opening, often accompanied by political instability and conflict, and transformation from centrally planned to open, market-driven economies, a process that frequently brought devastating macroeconomic instability, plummeting growth rates, and spiraling poverty and inequality.

Previously able to rely on central transfers of resources and guaranteed markets for their goods, these economies were characterized by enormously inefficient resource allocations. Unlike other regions of the world, infrastructure provision under Soviet rule had been extremely equitable—almost everyone had access to electricity and other basic services—but extremely inefficient. Now with transition coinciding with the global shift to market-oriented utility provision, the former Soviet economies naturally became the new testing ground for reform. The IFIs, as they assisted countries with reform programs focused on fiscal discipline and trade liberalization, placed substantial emphasis on increasing efficiency, eliminating losses, and introducing cost recovery in utility infrastructure. The electricity sector, given its size and importance to the fiscal budget, was a key contributor to the nonpayment problem, and it was inevitably among the first sectors to come under the spotlight.

Europe and Central Asia's Challenges Were Unique

The starting point of reform in ECA, and the challenges following the collapse of socialism, made reform in this region uniquely challenging. Incomes were higher than those in developing countries in other parts of the world, except Latin America. And other human development indicators—infant mortality, illiteracy, access to basic infrastructure, and progress toward the Millennium Development Goals—were better (table 1.1).

Infrastructure was also far more developed than in many parts of the world. The socialist legacy was publicly owned and vertically integrated, and its highly centralized power infrastructure was designed to provide reliable electricity to all households at little or no cost. Crucially, access to electricity was and remains substantially higher than in other regions

Table 1.1. ECA's Generally Higher Incomes and Better Human Development Indicators

	GNI per capita World Bank, Atlas method (dollars)	Adult literacy rate (2002)		Infant mortality	Access to an improved water source: percent of population	GDP per unit of energy use: purchasing power parity, dollars per kg oil equivalent
	(2003)	M	F	rate (2003)	(2002)	(2000)
East Asia and Pacific	1,070	90	86	32	78	4.6
Europe and Central Asia	2,580	98	96	29	91	2.5
Latin America and the Caribbean	3,280	86	88	28	89	6.1
Middle East and North Africa	2,390	82	61	43	88	3.5
South Asia	510	73	44	66	84	5.1
Sub-Saharan Africa	500	71	58	101	58	2.8

Source: World Bank 2005b.
Note: GDP is gross domestic product, GNI is gross national income.

with similar incomes (table 1.2), particularly for rural areas.[1] In urban areas, heating and often domestic hot water were also part of the cradle-to-grave centrally planned system.

But central planning also led to an inefficient and overdeveloped energy sector (figure 1.1), and with energy prices well below international prices, consumers enjoyed extremely low, nominal bills. Unsurprisingly, energy consumption levels were high.

Table 1.2. Access to Power Is Higher in ECA
(percent of households with electricity connections, 2000)

Region	Total (percent)	Urban (percent)	Rural (percent)	GDP per capita (percent)
East Asia	87	99	81	888
Europe and Central Asia[a]	99	100	97	1,998
Latin America	87	98	52	3,888
Middle East and North Africa	90	99	79	2,304
South Asia	41	68	30	441
Sub-Saharan Africa	23	51	8	496
World	73	91	57	5,216

Source: International Energy Agency 2000; World Bank 2000c.
a. Figures for ECA derived by authors from household survey data.

Figure 1.1. Energy Efficiency Is Lower in ECA

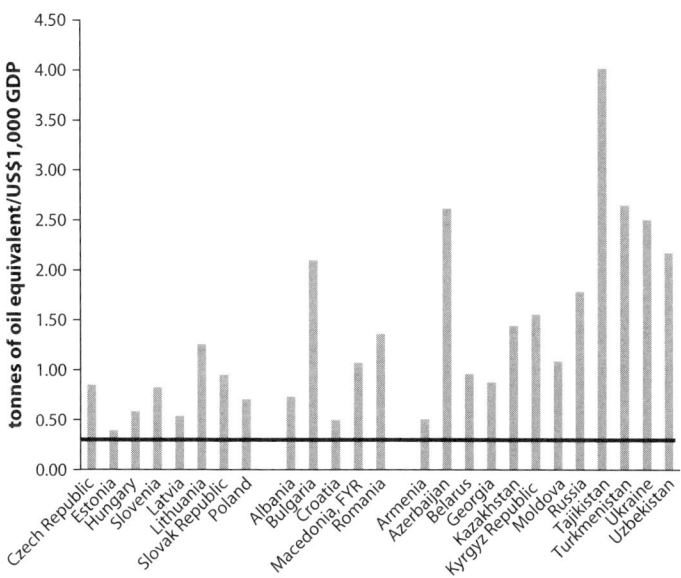

Source: EBRD (2001).
Note: The horizontal line indicates energy intensity of the United States. Using purchasing power parity (PPP)-cor-
rected results would yield similar results with a smaller gap with the U.S. Countries are in geographical groupings:
Central and Eastern Europe; Southeastern Europe; and Central Asia and the Caucasus.

The Onset of Crisis

For most transition countries, the early 1990s were years of economic
upheaval. Although the picture varies by subregion and country, depending on political stability and energy resource endowments, gross domestic
product (GDP) and real wages plummeted across the region while
inflation and fiscal deficits soared. With the end of central transfers and
associated price distortions, the former Soviet economies were faced with
skyrocketing market prices for fuel. Combined with low revenues from
customers, this meant that utilities, particularly in the electricity sector,
had to be supported by governments through indirect subsidies, cross-
subsidies, barter trading, and accumulations of arrears—a combination of
fiscal and "quasi-fiscal" transfers.[2] To absorb the costs of utility support,
governments were forced to run large deficits and accumulate foreign
debt. In some places, the energy sector deficit was one of the largest items
in the budget deficit, estimated at 11 percent of GDP in Armenia and
5 percent in Moldova.[3]

For economies already suffering painful transitions, subsidies on this scale were a further impediment to fiscal stability and recovery. The opportunity costs of such substantial transfers were enormous, and as money was funneled to support utilities, public spending on health and education fell dramatically. A decade after the onset of transition, Azerbaijan, Georgia, and Moldova's health spending was less than a quarter of what it was in the early 1990s. Armenia's education spending was a mere one-sixth of its level in the early 1990s, and Azerbaijan's one-third.[4]

Despite continuing government support to the power sector, utilities suffered significant financial losses and asset depreciation, and maintenance was neglected. Crumbling systems led to drastic declines in the quality and reliability of service delivery, with many consumers receiving electricity for only a few hours a day. The energy crises that emerged across the region, and the severe limitations they imposed on day-to-day economic activity, compounded the effects of economic collapse and held back recovery.

The Promise of Reform

The need to solve these energy crises and rebalance government expenditures made reform of the power sector an urgent issue for governments and donor institutions. The approach of the IFIs is crystallized in the World Bank's 1998 ECA energy sector strategy.[5] Formerly vertically integrated utilities would be unbundled into separately managed companies, and the sector would be deregulated, liberalized, and in many cases privatized. Prices would be raised to cost-recovery levels, to be enforced by metering and by cutting off nonpaying customers. Governments would establish predictable and transparent regulations, introduce competition in generation and distribution, sell industrial assets to private strategic investors, and improve the transparency of their financial flows by converting hidden budget support for utilities to explicit transfers. Donors in turn would provide funding to improve energy efficiency and advice on how to alleviate the impact of rising prices on poor households through means-tested transfers and tariff-based subsidies (table 1.3).

Unlike in Latin America and Africa, where reforms aimed at increasing access to electricity (equity), access in ECA was already almost universal. The major objectives of reform were thus to stop service quality deterioration and increase efficiency to improve the financial viability of the sector and reduce fiscal burdens (table 1.4).

Table 1.3. Components of Energy Sector Reform as Promoted by the World Bank in 1998

Category	Components
Demonopolization and regulation	• Unbundling vertically integrated monopolies to increase competition among energy producers and suppliers • Privatizing companies in competitive segments of the industry, shifting the role of the state from owner to regulator, promoting entry by foreign investors • Establishing liberalized and transparent markets for energy • Increasing autonomy, professionalism, and transparency of regulatory bodies
Prices and fiscal policy	• Setting prices at levels to ensure cost recovery and promote efficiency • Introducing taxes to compensate for negative externalities of energy production and consumption • Strengthening discipline in collection of payments (cutting off nonpaying customers, eliminating noncash payment methods) • Eliminating production subsidies, closing uneconomic energy production facilities
Foreign trade	• Opening domestic energy markets to external competition • Eliminating export taxes on fuels and electricity • Strengthening institutional framework for regional trading • Facilitating construction or rehabilitation of transnational energy connections
Investment policy	• Relying on energy companies (rather than budgetary resources) to mobilize investment funds in energy subsectors • Supporting investments in energy efficiency and the use of renewable energy resources • Providing information and risk mitigation to foreign investors to increase flows of foreign direct investment to the energy sector
Social protection	• Facilitating the shedding or redeployment of surplus labor and strengthening social safety net for the unemployed • Transferring social service functions from enterprises to local governments • Supporting poor urban and rural households through lifeline tariffs or means-tested subsidies
Environmental protection	• Supporting sectoral environmental assessments • Introducing emission norms for existing facilities • Analyzing environmental impact of new investments • Facilitating the mainstreaming of environmentally friendly technologies

Source: Adapted from World Bank (1998).

Table 1.4. Reform Goals and Indicators in ECA: Improved Service Quality, Resource Efficiency, and Fiscal Balances

Stakeholder	Outcome objective	Outcome indicator	Examples
Consumers	Improved service quality	• Reduced number of outages	• System average interruption frequency index
		• Frequency and voltage stability	• Number of deviations from established standards
Power sector (utilities)	Improved resource efficiency	• Increased revenue and collections	• Rise in electricity billed as percentage of net supply; rise in collections as percentage of billings.
		• Reduced cost of supply	• Reduction in cost of generation (dollars per KWh)
		• Improved energy efficiency	• Reduction in fuel use per KWh of electricity produced
		• Reduced losses	• Percent reduction (KWh lost per net KWh generated)
		• Improved operational efficiency	Rise in sales per employee; rise in consumers served per employee
Government	Increased financial independence	• Increased sector investment (third party)	• Percent increase in investment in generation, distribution, or transmission
		• Reduced sector financial deficit	• Percent decline in sector financial deficit expressed as a share of GDP

Source: Authors, based on reviews of project documents.

In a perfectly competitive economy, trade-offs between equity and efficiency take place along a production frontier. The objective of infrastructure reform is a function of the starting point of the reforming country within the production frontier and the type of reforms carried out.[6] In principle, ECA economies are well inside the production frontier: their power sectors were very equitable under the socialist system, but highly inefficient. With a balanced reform strategy, ECA countries could move outward toward the production frontier by improving efficiency without necessarily sacrificing equity.[7]

Improving cost recovery by increasing tariffs would create a financially sustainable power sector, freeing public resources for more productive investments (including in the social sector), and improved fiscal balances would lead to macroeconomic stability. Efficiently operated utilities would also mean better consumer service and environmental benefits from improved energy efficiency and investments in environmentally friendly technology. Lower emissions would lead to better ambient air quality and better health outcomes for the local population. Consumers would suffer because they would pay more for their electricity, but they would ultimately gain from improved service quality and macroeconomic stability.

For the poorest consumers, who have greatest difficulties paying and often the least access to substitutes, the impact would be greater, and the hardships particularly acute. But as with all reforms that generate an aggregate increase in welfare and an uneven distribution impact, the losers can be compensated. This means that it is particularly important for the government to make early decisions about whom to compensate and over what time horizon. According to public finance theory, and based on extensive scholarship that considers how to introduce cost recovery in infrastructure and other public services, the best solution is usually a lump-sum transfer, implemented as part of a social benefit transfer. Much reform in ECA has focused on moving from tariff-based subsidies—in the form of either across-the-board underpricing or lower tariffs for low volume consumers—to direct lump-sum transfers.[8]

The Problems of Reform

Though the necessity of reform was clear, there were problems and controversies from the outset, most obviously the backdrop of dramatically declining incomes across the region in the early stages of transition. The extraordinary upheaval of the move to market economies created enormous hardship and took a huge toll on standards of living. From 1991 to 1996, real incomes dropped by 14 percent a year, with only slight improvement in the remaining years of the 1990s. At the same time, ECA's climate limited how much people could cut back on energy expenditures. Winter temperatures can drop below −20° Celsius, and the heating season lasts on average five to seven months.[9] Households spend a large share of their incomes on energy for heat, and access to energy is a matter of survival. The legacy of free access and a sense of entitlement ensured controversy for any intervention to improve cost recovery.

Despite the considerable promise of reform, implementation soon proved more difficult than anticipated. Governments were slow to adopt reforms and many introduced parts of the package selectively (annex 1). In large part these differences were based on domestic political and economic conditions. While the movement to reform began in the early 1990s, for many countries, particularly those of the former Soviet Union, privatization started much later, if ever. Potential investors tended to be multinational companies, often based in the West, in search of new destinations for investments in the bull market of the 1990s. Inside the reforming countries, foreign ownership of utilities was widely viewed with suspicion, compounded by resentment over paying for a service that, for political reasons, used to be provided by the state at minimal cost. Many countries were highly ambivalent about reform, and though some chose to open the power sector to foreign investment, others considered generation assets as strategic and retained public ownership. Partial reforms were common, and progress was often the result of external pressure from donors, particularly for small, energy-poor countries such as Armenia and Moldova.

This ambivalence about reform can be traced in part to the mismatch between benefits and costs (table 1.5). While the costs are immediate, concentrated on a few groups, and highly tangible, the benefits take longer to accrue, even for governments and utilities. The fiscal benefits, one of the primary motives for reform, were slow to materialize and difficult to measure because of delays caused by institutions with vested interests, the appearance of formerly hidden transfers on the government's books, and expenditures on social transfers required to mitigate the impact of reform.[10] There was typically no systematic methodology to track and measure the fiscal benefits, and governments that should have been embracing reforms for fiscal benefits were not always doing so.

The picture was also ambiguous on the utility side, with strategic investors finding it difficult to recover costs in the face of fierce resistance from consumers unaccustomed to paying. The talents of enterprising (and desperate) consumers in tampering with meters and running dangerous illegal electricity connections from low-voltage cables made enforcement extraordinarily difficult. In the late 1990s, private investment in the sector fell steadily, while private operators, embroiled in contractual disputes,

Table 1.5. The Timing of Costs and Benefits Are Often Mismatched

Institutions	Costs (Usually immediate)	Benefits (Usually take time)
Government	Loss of control and rent-seeking opportunities	Improved fiscal balance
Utility	Loss of public financing	Financial sustainability and profit
Consumers	Increasing tariffs and disconnections	Improved service quality

Source: Authors.

withdrew or threatened to withdraw. In Kazakhstan, Belgian investor Tractabel walked out after tariff disputes with the government. In Moldova, the state initiated a lawsuit against Spanish investor Union Fenosa, arguing that the privatization process was flawed. In Georgia, U.S.-based AES Corporation described its purchase of the Tbilisi distribution company as a mistake and in 2003 sold its stake in the company to Russian interests after experiencing sustained losses. The growing ambivalence of companies toward the region and the profound changes in the world economy after 2000 resulted in a scarcity of strategic investors willing to pump the needed money into the sector, a scarcity reversed only recently in parts of the region.

Rising Prices, Rising Opposition

Perhaps the most immediate and visible effects of reform were rising energy prices and their impact on consumers, particularly the poor. Between 1991 and 2000, the price of electricity jumped by an average of 177 percent in real terms throughout ECA.[11] Universal access at little or no cost under socialism was clearly unsustainable, but it took time for consumers to adjust to the idea that services once provided for free must now be paid for. Cost recovery, in the form of improving collections and increasing tariffs, was immediate and visible. But the benefits for households—a desperately needed reliable supply of electricity and a chance for macroeconomic stability—would take longer to accrue, and economic growth would benefit the populace only through less visible second-order effects. In the interim, rising energy prices clashed with falling incomes, rising income polarization, and alarming levels of urban poverty (figure 1.2). And poor data made it difficult to assess whether the poorest were being adequately compensated.

As the 1990s progressed, the emerging picture in many places was of incomplete reforms and ambiguous results and benefits. Across ECA the picture varied, with reform in the Baltics and some countries in Central Europe reasonably rapid and successful. Elsewhere, difficulties in identifying and communicating the benefits of reform made it all the more difficult for governments to credibly justify tariff increases. Amid the complicating factors, one certainty was emerging: public concern over the effects of rising prices and privatization on the poor was helping to create and sustain significant and organized constituencies that opposed reform. The increasing tendency to doubt the virtues of reform was fueled by external developments: high profile "failures" such as Russia and the

Figure 1.2. Rising Energy Prices Clashed with Falling Incomes in ECA, 1991–2000

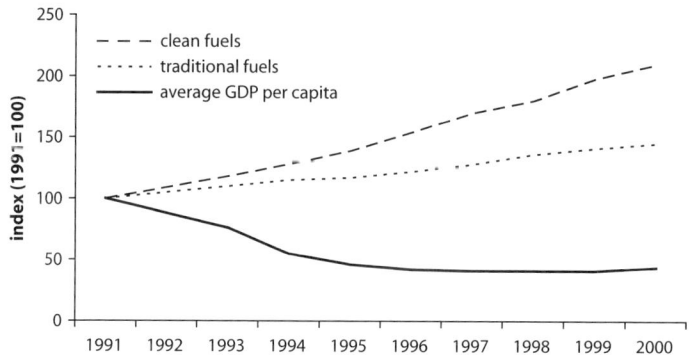

Source: Authors' calculations from International Energy Agency data and World Bank data. These figures may not be equal to the true resource cost because of the effect of subsidies (Lampietti and Meyer 2002).

Czech Republic's voucher privatizations and the backlash against the "shock therapy" of the early 1990s; the gaining momentum of the antiglobalization movement and its opposition to the market-driven Washington Consensus supposedly championed by the IFIs; and the emergence of widespread public campaigns against utility reforms—most notoriously the clash over the Cochabamba water utility in Bolivia in 2000. Although not necessarily backed by empirical evidence, a popular perception that privatization failed consumers, combined with domestic opposition to increased tariffs, further undermined confidence in reform among people in ECA and the governments who needed their support.

Much progress had been made in some countries—Bulgaria, Hungary, and Poland, among others—but in many countries by the late 1990s, particularly in the former Soviet Union, reform and privatization were perceived by many to have failed to live up to initial expectations. The bursting of the privatization bubble and the deviation of actual reform outcomes from intended outcomes posed a growing threat to the continuation of reform. Countries that had entered the process were reluctant to push for further reforms, especially tariff increases. Certain governments were distancing themselves from reform, some overruling tariff increases set by independent regulators, and undermining the efforts of utility operators to turn the sector around. Some governments, such as Moldova, even threatened to backslide and reverse reforms, while countries that had not yet reformed their utilities were concerned about the social and political fallout of doing so.

Notes

1. Komives, Whittington, and Wu (2001); Clarke and Wallsten (2002).

2. Nonmonetary government support, often termed quasi-fiscal transfers, includes subsidized supplies, tax exemptions, or bartering of services with other state enterprises, which does not therefore appear in the state budget as a transfer.

3. Energy sector constituting electricity and gas. The Armenia figure is for 1995, the Moldova figure for 1999 (Sargsyan, Balabanyan, and Hankinson 2005; IMF 2001b).

4. As shares of GDP, total public expenditures on education, health, and social assistance and welfare remained stable or fell (Public Expenditure Database, World Bank 2002).

5. World Bank (1998).

6. Birdsall and Nellis (2003, 2005).

7. Here the word "equity" is used in the same sense as in the *World Bank's World Development Report 2006: Equity and Development*, as ensuring that individuals have equal opportunities (in this case access to electricity) and are spared from extreme deprivation in outcomes (World Bank 2006).

8. For a more detailed discussion of the different types of subsidy available see Komives and others (2005), chapter 2, "A Typology of Consumer Utility Subsidies."

9. Exposure of populations to extreme temperatures was in some cases exacerbated by Soviet planning policies, which encouraged settlement in areas with cold climates, such as Siberia (Hill and Gaddy 2003).

10. The fiscal deficit is the difference between revenues and expenditures as recorded in the official government budget. In addition to fiscal deficit, public finance analysis takes into account government obligations that are not reflected in the budget, but result from explicit or implicit government liabilities outside the budget framework. When a utility is publicly owned, the government receives taxes and dividends from the utility and provides explicit and implicit subsidies, many of them not transparent. They could be explicitly recorded in legal documents or result implicitly from the logic of political events, institutional rules, or social obligations of the government as understood by the public. Untangling these financial flows requires detailed systematic data on financial flows that are not readily available. The data and analysis of the electricity sector fiscal and quasi-fiscal deficits are available for the countries for which the International Monetary Fund or the World Bank undertook detailed studies, such as Armenia, Romania, and Russia (Petri, Taube, and Tsyvinski 2002; Frienkman, Gyulumyan, and Kyurumyan 2003; Saavalainen and ten Berge 2003). Some

evidence also exists for Georgia and Moldova, but no systematic methodology or time series data have been available to date.

11. These data cover Armenia, Azerbaijan, Estonia, Georgia, Kazakhstan, the Kyrgyz Republic, Latvia, Lithuania, Moldova, Tajikistan, and Uzbekistan (Lampietti and Meyer 2002).

CHAPTER 2

Using Poverty and Social Impact Analysis to Assess the Distributional Impact of Power Sector Reforms

A sustainable power sector rested on raising tariffs from below cost to levels where utilities could recover costs. But tariff increases have contributed most to the widespread mobilization of opposition to utility reform, even where reform has dramatically improved access. The justification for higher tariffs can be based on pro-poor arguments. When tariffs are below cost recovery, the government budget subsidizes the electricity consumption of all members of society, poor and nonpoor. This subsidy often comes at the expense of macroeconomic stability and much-needed investments in other sectors, including the social sector, that more directly benefit the poor. And since people who are better off generally consume more electricity, they capture the bulk of the subsidy in absolute terms. Subsidized provision of electricity to all consumers, as well as being extremely costly and encouraging inefficient use of electricity, is thus socially regressive, and subsidies are commonly criticized for being unpredictable, unsustainable, and unaffordable.[1]

But removing across-the-board subsidies presents its own problems. As tariffs increase to cost-recovery levels, either the consumption of electricity must decrease, or the share of household income spent on electricity must increase, or both. It may be that better-off households spend more on electricity in absolute terms, but the share of income spent on electricity is usually larger for the poor, so a tariff increase will affect them more. When the ratio of income spent on energy exceeds a certain threshold, households are in danger of becoming energy poor. Once they have cut back on all inessential electricity consumption, they must sacrifice the consumption of other goods to satisfy their basic energy needs.

To prevent the potentially substantial welfare loss that results from crossing this threshold, economists usually favor lump-sum transfers to the most vulnerable consumers. But whether tariff-based subsidies or lump-sum transfers are more effective and efficient in assisting the poor rests on answers to questions about access to subsidized utilities, household consumption, and where transfers go. To meaningfully analyze the distributional impact of reform and how this can be improved through better policies requires reliable information—on access to energy, income, energy consumption, basic minimum needs, coping mechanisms used by households when energy prices increase, and how effective different subsidy or lump-sum transfer systems are at mitigating the impact of reform on the most vulnerable. Based on an empirical understanding of budget shares spent on electricity, the methodology behind the studies in this book can help answer these questions, showing who the winners and losers of reform are and how the losers can be compensated.

Why These Studies?

When the studies were conceived, it was clear that the intended outcomes of energy sector reform were not materializing as quickly as expected. The fiscal benefits of reform were obscured in a haze of indirect government subsidies. Utilities were charging more for electricity, but the expected returns on their investments were elusive as consumers resisted tariff increases. And the poor were suffering, in many cases more than expected. While donors pointed to net welfare improvements resulting from reform,[2] popular protests, politicians, and opposition groups attested that the more immediate effects—increasing tariffs, collections, and disconnections—were felt far more strongly. Reform was proving tough on consumers and on governments trying to administer it. The political consequences of rising prices, combined with the less-than-perfect outcomes for governments and utilities, threatened to bring reform to a halt and deter other countries from reform altogether.

In the absence of routine attempts to quantify the distributional impacts of large sectoral reforms, the impacts were not well understood. The donor community and governments lacked an empirical understanding, either ex ante or ex post, of what was going on. In theory, means could be found to compensate the losers, but the losers had to be identified. If the poor had gained from better service, but lost to increasing prices, collections, and disconnections, what was the net effect on their welfare? How could the design of reform be improved to soften the blow? A systematic, analytical approach was needed to shine new light on the empirical effects on the poor and to differentiate between the reality and perception of reforms.

These studies identified the possibilities offered by the available data for empirically identifying the direction and magnitude of the impact of electricity reforms on welfare distribution—and the potential of policy analysis tools for producing a picture of household behavior under reform.

Who Are the Stakeholders of Reform?

The aim of these studies was to improve understanding of the distributional impact of reform on primary stakeholders, focusing on the poor. The primary stakeholders are utilities, government, and consumers.[3] For utilities, reforms aim at distancing them from political control and introducing profit as an incentive for greater efficiency. To minimize costs utilities will make more efficient resource allocations, while improving cost recovery through tariff increases allows them to invest in maintenance and repair. The net effect is a more efficient sector that is financially sustainable and delivers a better service to consumers. Furthermore, establishing a regulatory body that is independent from the government can improve the situation of utilities since—in theory at least—they are no longer subject to political pressures to provide cheap electricity.

Governments will benefit from reduced sector liabilities and fewer indirect transfers. This promotes a stable macroeconomic environment, which helps economic growth and allows public investments in other priority sectors. Indirect government transfers are converted to quantifiable subsidies, which improves government record keeping and budgeting. Privatization allows governments to get the utilities off their books entirely. An independent regulator, setting service quality standards and regulating tariff increases, allows the government to distance itself from utility price increases, sending a signal to private investors that the government is serious about reform and about improving the investment climate.

Consumers can be divided according to their income level, whether they are urban or rural, by geographical region, or by different types of fuel users.[4] In general, consumers in ECA are expected to lose from increases in tariffs and collections, but gain from improvements in service quality and availability, and ultimately from macroeconomic stability and higher social sector spending. The magnitude of any one of these benefits can be great and depends on several factors.

Households that had reliable service and consumed a lot of electricity but did not pay their bills will lose because service quality improvements will be minimal, but their costs will increase. Households that had illegal connections that are now curtailed will lose, too. Households that previously faced electricity rationing or voltage fluctuations that ruined their appliances will also lose from price increases, but they will gain considerably from improvements in service quality and supply. They will see a net gain in welfare if improvements in service quality are sufficient to make up for the welfare loss incurred as a result of tariff increases. For households that receive government benefits, the impact of increasing tariffs will be greatly softened. Households able to switch away from electricity to cheaper fuels will also be better off compared to those more dependent on electricity (table 2.1).

The Theoretical Basis

The theoretical framework of these studies lies in social cost-benefit analysis. This is part of the range of elements in poverty and social impact analyses (PSIAs), along with analyzing stakeholders and the institutions implementing the reform, and identifying channels that transmit impacts and the risks to the reform.[5] The PSIAs examine empirical data on the impact of reforms and approximate net welfare changes, in this case as they accrue to the primary stakeholders, households, government, and utilities.

The vast scale of utility sector reform means that household consumers are affected both directly and indirectly. Consumers are directly affected by improvements in service quality and increased prices. But electricity is also an important input for producing goods. A reform that substantially affects the availability of electricity or increases the cost can profoundly influence the cost of the basic consumption basket. Electricity reform will also have macroeconomic effects, an important determinant of the welfare of all groups in society. And as fiscal deficits go down, the impact on growth and on other areas of government spending should be positive, which will also improve the welfare of the poor.[6]

Table 2.1. Winners and Losers from Reform—A Typology of Consumers

Consumer characteristics before reform	Welfare impact of reform			
	Improved service reliability	Improved service quality	Higher tariffs	Greater payment discipline
Enjoys reliable and good quality service				
Little or no payment	No impact	No impact	Very negative	Very negative
Eligible for social benefits	No impact	No impact	Negative, depends on level of assistance	Negative
Access to cheaper substitutes	No impact	No impact	Negative, depends on ability to substitute	Negative
Limited access to reliable and good quality electricity				
Little or no payment	Positive	Positive	Very negative	Very negative
Eligible for payment social benefits	Positive	Positive	Negative, depends on level of assistance	Negative
Access to cheaper substitutes	Positive	Positive	Negative, depends on ability to substitute	Negative
Access to unreliable and low quality electricity				
Little or no payment	Very positive	Very positive	Very negative	Very negative
Eligible for social benefits	Very positive	Very positive	Negative, depends on level of assistance	Negative
Access to cheaper substitutes	Very positive	Very positive	Negative, depends on ability to substitute	Negative
Consumes free electricity from illegal connection, no payment	Slightly positive	Slightly positive	No impact	Very negative
No access to electricity, consumes alternate fuels	No impact	No impact	Limited impact, but can be priced out	No impact

Source: Authors.

To quantify all these effects requires general equilibrium analysis, using a computable general equilibrium (CGE) model. This enables a broad assessment of the net effects of reform on the entire economy. As well as the direct or first-order effects of sector reform on the consumption of electricity, CGE models capture the indirect or second-order effects. They can quantify changes in the consumer basket as a result of changes in input prices of different goods. They also allow simulation of the macroeconomic effect of reforms, factoring in such effects as reduced fiscal deficits.[7]

But CGE models have drawbacks that make them inappropriate for the studies here. They require a substantial volume of micro- and macroeconomic data that must be entered into a Social Account Matrix, data that take time to collect and that may not be uniformly reliable. They also require a much larger number of assumptions about relationships between variables, with the reliability of the entire model resting on the accuracy of these assumptions. For a study that aims to bring clarity to policy debates, the complexity of the CGE process, and its inaccessibility, are major disadvantages.

Other analytical tools can give valuable information on the first-order effects of reform but require fewer resources than a CGE model.[8] The studies here looked primarily at trends in the share of monthly household expenditures on energy (budget shares), comparing budget shares across income groups. Welfare analysis looks at which groups are seeing benefits from a policy. Changes to consumer surplus provide a measure of changes in welfare (in the Azerbaijan case study and in looking at energy sector reform across ECA in chapter 9). Proxy determinants, such as the incidence of disease, can be used for the key nonmonetary dimensions of well-being (to a limited extent again in chapter 9, on the environmental impact of reform). And contingent valuation can be used to infer the willingness to pay to assess demand for heat (chapter 8).[9]

Basing the analysis on one or more of these approaches provides a fairly straightforward process that gives empirically useful results, at the same time using fewer assumptions and more readily available data than a CGE model. A simpler procedure, it uses methods and produces results that are more easily explained to stakeholders—who can participate in the analysis and use the tool in future analyses. Given that the PSIAs were intended to tap local capacity, this also made the simpler method attractive. The studies paint a picture of how a small number of variables, such as price and availability of electricity, affect the welfare of a stakeholder group, and they do this in a way widely understood by policy makers.

The studies do not capture the second-order effects of reform that would be seen in a CGE model, but they do examine some of the links that contribute to a more comprehensive view of reform than would come from a partial equilibrium analysis. They may not simulate the effects on the entire economy, but they do take into account the effects on stakeholders other than consumers, and they look explicitly at the link between household consumption and utility revenue. In Armenia, the study used household data to show, in the short term, that although revenues from residential consumption should have increased by 16 percent with the tariff increase, collection rates fell by almost 10 percentage points, meaning that residential revenues increased by only 6 percent. The studies also consider the impact on the fiscal deficit, a prime motive for reform. In addition to examining the effectiveness of the social assistance system, the Georgia study found that the government was spending a lot of resources on a subsidy captured largely by higher income households, and that subsidies for gas consumption were increasing, encouraging people to switch to gas. It also found that the money that the government was saving on subsidized electricity was not being channeled into social spending.

The studies also look at some of the secondary effects of reform, including the social and environmental costs of households switching fuels as a result of changes in relative fuel prices, such as the time taken to gather wood and the indoor air pollution associated with burning these traditional fuels. The Azerbaijan study considered the economic gains from improved access to electricity in the agroprocessing industry, which will gain from a reliable energy supply through additional revenues and cost savings. It also examined how projected income growth could alleviate the welfare impact of increased tariffs.

Welfare Indicators and How to Measure Them

To assess the distributional impact of electricity reform measures—tariff increases, greater collections of tariffs, and changes in service quality and availability—the studies built a comparison of budget shares spent on different types of energy across a specified period of reform. This comparison required extremely reliable information on a specific set of diagnostic welfare indicators that included household income and expenditure, electricity consumption, and absolute electricity expenditure. The budget share analysis was supplemented by information on service availability and quality.

These welfare indicators can be compared across different groups of consumers—poor and nonpoor, urban and rural, those with access to substitutes and those without—to compare how reform affects different stakeholders. For example, how do different quintiles of the population respond to changes in price? This information enabled comparisons of the net welfare effects of reform on these stakeholders. (For more information on the methods and sampling techniques, see http://wbln0018.worldbank. org/esmap/site.nsf/pages/Flagship_2006).

Qualitative Analysis

Qualitative analysis involves consulting a variety of stakeholders, including representatives from consumer groups and utility companies, government, the private sector, regulators, and households from different socioeconomic strata and with access to different substitutes to electricity to obtain their views and experiences of reform. The analysis draws on focus group discussions and in-depth interviews of key informants.

Qualitative analysis complements the quantitative analysis and has two important uses. When it is carried out before the quantitative analysis (the preferred approach in these studies), it can generate testable hypotheses about behavior in response to higher tariffs and thus inform the design of the quantitative survey. In Azerbaijan, for example, the qualitative analysis helped in developing the typology of households using different fuels and in designing the survey questions to capture and measure behavior. It also provided important knowledge on the kinds of appliances used by households. The focus groups also helped identify the aspects of the reform program that people are particularly concerned about and thus what questions the quantitative surveys should include.

In Georgia and Moldova, the survey data were already available for analysis from earlier surveys, and the qualitative analysis came second. This approach has advantages when the qualitative data can help shed light on otherwise opaque quantitative findings and paint a more complete picture.

In the Moldova study, the qualitative analysis confirmed the validity of the findings on household behavior gleaned from the quantitative analysis. The quantitative analysis had suggested that the poor were generally doing better than they had been when reform was introduced: electricity consumption was rising along with incomes, and electricity expenditures represented a decreasing share of the household budget. But the qualitative analysis, from focus groups with a representative cross-section of households, revealed that despite these improvements, people still faced serious hardships, and enforcement of payment was resented particularly by the poor. Consumers were compelled to take extreme measures, such as unplugging their refrigerators for days at a time, to keep their electricity consumption down to an affordable bare minimum. While the data revealed that average electricity consumption had increased by 10 percent, from 50 KWh to 55 KWh, the focus groups revealed that this improvement was imperceptibly small.

In Georgia, the quantitative data showed that households in Tbilisi were maintaining fairly stable energy expenditures and consumption levels despite tariff increases. This implied that they were replacing electricity with less expensive fuels. The focus groups confirmed that households with access to gas preferred to use gas when possible, since it was cheaper than electricity and cleaner and more convenient than other substitutes, such as kerosene and wood. Conversely, households without access to gas were using kerosene or wood for heating and cooking and desperately wanted access to gas—feeding into a key recommendation at the time to subsidize the extension of the gas network to poor neighborhoods.

Quantitative Analysis

Several methods were used to obtain data for the quantitative analysis of welfare indicators. The bulk of the data for the quantitative analysis came from household budget surveys (HBSs) containing data on general household expenditure (as a proxy for household income) and energy consumption and expenditures. First explored was the possibility of using an existing data set, such as a Living Standard Measurement Study (LSMS) or HBS. If no appropriate data existed, a new primary data collection exercise was initiated, keeping the sampling frame and parts of the questionnaire consistent with the most recent HBS or LSMS—to ensure that the PSIA work was consistent with the broader poverty assessment.

A typical HBS includes a household roster to show the size of the household, and questions on household monthly income and expenditure, expenditure on electricity, and, where possible, fuels that the household uses as substitutes for electricity.[10] Although HBS or LSMS data were used where already available, the surveys designed especially for the studies contain far more detailed questions about the number of fuels used and what they are used for, as well as questions about the household's attitudes and perceptions of electricity sector reform and tariff increases. Ideally, the households surveyed are the same before and after reform, to determine the effect of reform on them and to obtain a dynamic picture of household welfare.

Generating Better Data

HBSs are widely used as a tool to analyze poverty. The PSIAs for this book used a key empirical enhancement, however, by collecting the billing and payment records from utilities for the same households covered in the HBS. The billing and payment records were then merged with the HBS to correlate the data for the same household in the same time period. This is a time-consuming process that requires complicated database manipulation, but the information on household consumption produces important results.

Merging the two data sources provides external validation of the self-reported electricity expenditure data in the HBS. While questions on electricity consumption and expenditures can be included as part of traditional poverty monitoring surveys such as the HBS, self-reported electricity and energy expenditure data collected in these surveys are notoriously unreliable. The results are often confounded by recall error, under- and overreporting, and the presence of arrears, making it almost impossible to identify current and historical consumption. The potential disparity between HBS figures and the data from utility records is illustrated in the Georgia study, which displayed significant discrepancies between reported and utility data. Payments reported in the HBS were consistently higher than those recorded by the utility, a finding that might be attributed to corruption, with households paying more to meter readers than meter readers transfer to the utility, or to recall error, which is easily explained if the households are reporting bills received rather than payments made. Conversely, in the Moldova study, the two sets of data from the HBS and the utility records were highly correlated, increasing confidence in the HBS data.

Matching household survey data on income with household data from the utility on electricity billing, consumption, and payment allows a much more reliable and sophisticated analysis of residential demand, of who is getting what, and how much they are consuming. It provides empirical data on the distributional impact of price changes, service quality improvements, and other reform impacts, since the data can be tracked over time for the same household. It permits a reliable comparison of how much households with different characteristics are spending on electricity—how much the poor and the nonpoor consume and differences in consumption between rural and urban households. And it makes it possible to disaggregate the impact of rising tariffs from rising collection rates, since it can be seen who is paying bills, who is accumulating arrears, and how much they are accumulating—allowing a determination of whether price changes have contributed to thefts of electricity or to changes in consumption, or to both. Knowing which households are accumulating arrears—whether they are poor or nonpoor, and, therefore, whether nonpayment is due to free-riding or affordability—also tells whether policy makers need to focus on improving enforcement or social transfers.

The Georgia case is an example of the kind of insight that reliable consumption and expenditure data can give. The utility data allow a careful examination of household electricity consumption patterns over the past three years. While prices increased, mean household electricity consumption remained constant at about 125 KWh per month. This finding has two implications for policy makers. First, current consumption levels are extremely low. Basic minimum consumption is likely to be approximately 125 KWh per month, roughly enough electricity to power a refrigerator and three incandescent lightbulbs. Second, demand in Tbilisi, where service has been quite reliable for the past few years, remains constant despite price increases, suggesting inelastic demand and large welfare losses from future price increases.

The Moldova study showed changes in welfare indicators—including access to, consumption of, and expenditures on electricity—comparing poor with nonpoor households. The household survey also highlighted coping mechanisms induced by increasing tariffs, consumer perceptions of reform, and the role of the social assistance compensation system. The analysis answered very specific but politically charged questions, such as whether the impact of reform was different for the poor and the nonpoor, and whether those served by the private operator were worse off.

Reliable data on consumption of electricity and substitutes also allow a more nuanced examination of alternative social protection measures, such as income transfers and lifeline tariffs, including their social and fiscal impacts. Knowing the income and consumption patterns of individual households made it possible to simulate how much they will receive under different transfer regimes, and how this compares with their welfare under a different mitigating strategy. It was also possible to see whether transfers are well targeted and how much different protection measures will cost the government. In the Georgia study (chapter 5), this insight enabled a comparison of the benefits and costs of the current social protection strategy with an alternative strategy devised by the authors, and the determination that the alternative strategy would be more effective at targeting poor households and would cost the government less.

The reliability and detail of the information created by merging the survey and the utility data also allowed the authors to build a demand model to simulate energy consumption at different levels of energy price and income. In the Georgia study, an electricity demand model demonstrated the minimum level of consumption below which demand is extremely inelastic. In Azerbaijan, the methodology was carried a step further; utility and survey data from several countries were combined to estimate an electricity demand model. This model was then used to predict the household consumption response and the welfare consequences of different potential rates and levels of tariff increase, which can help in designing reform policy.

Another innovation that allowed new empirical insights was a model for estimating energy and electricity demand. In the study on heating strategies for the urban poor in chapter 8, the authors developed a model for estimating household heat consumption.

Limitations of the Methodology

There are some important qualifications when considering the findings in these studies. Perhaps the most significant weakness, which this methodology shares with other techniques, is the inability to compare the findings on household impact of reform with the counterfactual—the situation households would find themselves in if reform had not taken place. To some extent, the situation of households before reform can be taken as an approximation for the counterfactual, since it can be assumed that this situation would have continued in the absence of reform. This is most effective when the same households are compared before and after reform (a panel study); only the study for Georgia had this data. In the Moldova study, the ability to compare households served by a private operator with those served by a public utility also goes some way toward a counterfactual.

Another potential weakness, already discussed, is the methodology's inability to model the second-order effects of reform on household consumers, effects that might be transmitted through various channels, including changes in prices, assets, access, employment, or transfers. For example, though the studies show the impact of rising prices and improving service quality, consumers will also be affected by improved macroeconomic stability resulting from the fiscal benefits of reform, and from economic growth resulting from increasing access to electricity (though these issues are mentioned in several of the studies, particularly for Georgia and Azerbaijan). This is not a major shortcoming, however, given the purpose of the studies: to find ways to smooth the transition to cost recovery through an understanding of the first-order effects of reform, rather than to provide a picture of the aggregate welfare impact on the economy as a whole.

The Advantages of PSIAs for Designing Reform

The PSIAs were valuable because they presented an opportunity to conduct a more robust empirical analysis of the social consequences, particularly those relating to poverty elements of the sector reform program. They generated specific analytical innovations and provided a critical emphasis on household behavior and choices, with the analysis contributing to new ideas on how to mitigate the negative impact of reform. PSIAs showed a more complete picture of winners and losers from reforms, and how to compensate the losers. Moreover, they can be used to draw policy recommendations for formulating less contentious reform in the future.

The combination of appropriate tools for analysis and creative use of data produced a unique story of the household level impact of reforms. The ex post analyses of Armenia, Georgia, and Moldova offer an important empirical record of how power sector reform affected the poor. This understanding gives guidance on how to modify the design of reform going forward, in the country in question and in similar cases in the region and beyond. Ex ante studies, of Azerbaijan and of heating strategies for the urban poor, also have clear implications for operational design. The simulation of how different reforms will affect welfare is invaluable in informing decisions on which policies to adopt.

The ability to map and simulate the effects of reform and the innovations in data use made the PSIAs a significant contribution to the tools for evaluating and designing policy reforms. But perhaps the most important contribution is that the process and findings of the studies encourage public discourse on the reforms. With these PSIAs, a working group of government, civil society, and nongovernment stakeholders can be brought in at the concept stage to participate in the analysis and discuss the findings. Largely through the collection of qualitative data and analysis of quantitative data, but also by forging new insights about the effects of reform, both the process and the findings of the studies can generate stakeholder dialogue and engagement.

Stakeholder engagement and public discourse can slow the process of reform. But in recent years they have become part of donor-funded projects and are expected to have a significant impact on different stakeholder groups as part of a more participatory approach to development. Experience from these and other studies demonstrated that time spent in stakeholder engagement and dialogue could not only help build consensus on reform, but also actually improve the design and outcome of reform, making it more sustainable. By enabling a better dialogue and understanding of the implications of reforms and the possible alternatives, the PSIAs could prevent questions from becoming politically fraught in the first place. In Moldova, for example, the PSIA brought into the open several unexpected findings that had been at the heart of the debate over the merits and costs of reform and privatization.

The open process of information gathering and discussion with stakeholders can thus foster local ownership of the studies and enhances the credibility of their findings. The involvement of stakeholders and local expertise in the design and execution can also contribute to building capacity at the local level, among consultants, working committees, and governments. Through their involvement they can do more in designing reform and in weighing the trade-offs. Since the goal is for countries eventually to carry out these studies independently, this is a significant contribution. And by emphasizing the importance of empirical analysis in designing and measuring the impact of reform, the studies also highlighted the importance of careful record keeping by the government and by the utility.

Since the first of these studies was commissioned, PSIAs have come to be regarded in the World Bank as best practice "to promote evidence-based policy choices and foster debate on policy reform options."[11] PSIAs are now meant to be embedded and mainstreamed in country work to improve policy design and the outcome and sustainability of reform. PSIAs are particularly important in the design of reforms that "are expected to have large distributional impacts, are prominent in governments policy agenda, and are likely to involved significant debates," all of which characterize power sector reform in ECA.[12]

Notes

1. For a more detailed discussion of the costs of across-the-board subsidies see Lovei and others (2000a) and Komives and others (2005).

2. World Bank (1998).

3. For a broader discussion of stakeholders and tools for their analysis, see www.worldbank.org/psia and World Bank (2003g). For a broader discussion of reforms of utility providers, see chapter 3 in Coudouel and Paternostro (2005).

4. For a more complete analysis of the range of consumer stakeholder groups, see Foster, Tiongson, and Laderichi (2005), pp. 84–88.

5. See World Bank (2003g).

6. The channels for reforms to affect different groups can be classified as employment, prices (production, consumption, and wages), access, assets, and transfers. For a discussion of these channels, see World Bank (2003g).

7. See, for example, Chisari, Estache, and Romero (1999).

8. For a review of the broad range of tools available to measure distributional impacts, ranging from simple incidence analysis to more complex models linking macroeconomic models with microsimulation, see Bourguignon and Pereira da Silva (2003), available at www.worldbank.org/psia.

9. For more on different techniques to measure or predict the impact of reform, see chapter 3 in Coudouel and Paternostro (2005), pp. 107–12.

10. See the HBS sample at http://wbln0018.worldbank.org/esmap/site.nsf/pages/ Flagship_2006.

11. Coudouel and Paternostro (2005), p. xi.

12. Coudouel and Paternostro (2005), p. xi. Other examples include trade, monetary, and land policy reform. For illustrations of the particular aspects of selected reforms, see Coudouel and Paternostro, eds. (2005). For examples of applications of the PSIA approach to other countries and sectors, see "A User's Guide to Poverty and Social Impact Analysis," at www.worldbank.org/psia.

PART 2

Case Studies

CHAPTER 3

Energy Reforms and Trends in Household Consumption

Between 1990 and 1997, per capita commercial energy consumption across the Europe and Central Asia (ECA) region fell by one-third.[1] Though much of this drop can be attributed to the collapse of industry, there also appears to have been a fundamental shift downward in residential energy consumption, attributable to the decline in subsidized infrastructure services, coupled with rising poverty and higher prices of basic goods and services. This chapter reviews energy sector reforms in ECA and changes in residential energy consumption over the past decade and a half.[2]

Patterns of Reform

Energy sector reform included unbundling, privatizing, establishing independent regulatory bodies, and improving cost recovery.[3] But reform has varied widely across the region, with differences in how reforms have been adopted and their success (table 3.1 and annex 1). These differences can be understood through the prism of political, institutional, and macroeconomic conditions, including energy endowments, accession to the European Union (EU), and the accumulation of energy-related debt.

Most Central and Eastern European countries made early progress in reforming the energy sector. They displayed better macroeconomic performance and provided a reasonably attractive environment for foreign investors. The prospect of EU accession and the need to conform with the EU directive on power reforms also provided the impetus for fast-paced reform, especially in developing a regulatory framework and unified gas and electricity markets.[4] Hungary was the leader in pursuing major electricity privatizations in the 1990s, with most electric utilities privatized and tariffs at the world market level.[5]

Countries in Southeastern Europe and the former Soviet Union were frequently characterized by war and civil unrest, volatile political conditions, destroyed physical infrastructure, risky investment climates, weak administrative capabilities, and low utility payments, resulting in extreme decapitalization of the sector.

Possessing energy resources could both ease and hurt progress with reform.[6] Energy-exporting countries such as Azerbaijan, Kazakhstan, and Turkmenistan gained from a change in their terms of trade during transition; they were able to export their energy resources at the higher world price, staving off fiscal crises. Energy-poor countries such as Armenia, Georgia, and Moldova lost due to dependency on unreliable and expensive external energy supplies. Forced to accumulate energy-related debt, they did not have the resources within the energy sector to mitigate the adverse social impact of reforms.

Table 3.1. Timeline of Reforms in the Electricity Sector in ECA

	Date of passage of energy law and creation of an independent regulator	Corporatization and unbundling	Privatization of distribution	Privatization generation
Armenia	1997	1997	2002	2002–03
Azerbaijan	(2006)	1998	2002 (management contract)	None
Georgia	1997	1999–2000	1998	2000
Hungary	1993–94	1993–94	1995	1996–97
Kazakhstan	1998–99	1996	1996, 1999	1996, 1999–2002
Moldova	1998	1997	1999	None
Poland	1997	1993	Ongoing	None[a]

Source: Adapted from Krishnaswamy and Stuggins (2003).
a. Except for new entry of private strategic investors.

But energy endowments can also constitute a barrier to reform when resource rents are appropriated by ruling elites.[7] Just as the symbiotic relationship between energy utilities and the government meant widespread corruption and rent-seeking in the pre-reform era, there are powerful motives for ruling elites to benefit from partial reforms in lucrative sectors by gaining control of the regulatory process and preventing the creation of a level playing field. By contrast, the need to reduce energy-related external debt has been a significant driver of reform in heavily indebted Armenia, Georgia, and Moldova,[8] and conditional lending by donors was used in an effort to promote reform (though the performance of conditionality has been mixed).[9]

Trends in Residential Electricity Consumption

The second half of the 1990s saw a gradual return to political stability, paving the way for economic reform programs, and most countries saw stabilization and even improvement in their economic situation by the end of the decade.[10] But the rocky transition of the 1990s took a heavy toll on living standards and equality, and high, if declining, levels of poverty still characterize much of the region (figure 3.1).

ECA households in the 1990s were spending 2–10 percent of their income on electricity.[11] The lowest 20 percent of the households, the poor, consistently spend a larger share of their income on electricity than the top (figure 3.2), suggesting that once a certain minimum level of consumption is reached, consumption becomes price inelastic.[12] If true, this analysis implies a greater proportionate welfare loss for the poor and a more active search for substitutes when tariffs are increased to cost-recovery levels.

Figure 3.1. Poverty in ECA Increased with the Transition

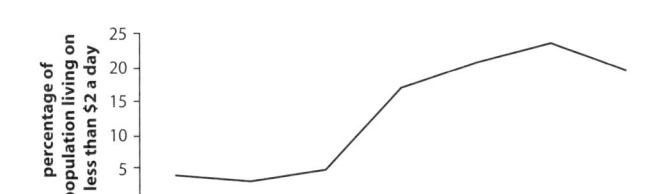

Source: World Bank (2005b).

Figure 3.2. The Poor Spend a Larger Share of Their Income on Electricity

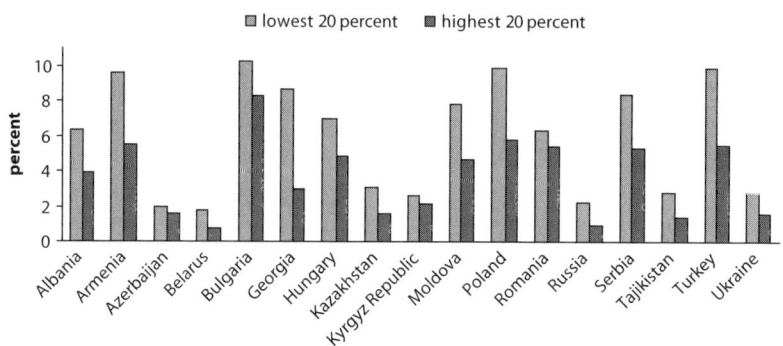

Source: Author's calculations based on household survey data from the World Bank ECA HBS database.
Note: Conditional on households reporting positive expenditures. Figures for Bulgaria and Tajikistan are for 2003.

That the poor spend a larger share of their income on electricity also implies that they use it for needs for which other fuels are poor substitutes, such as lighting, refrigeration, and television. Wood, kerosene, liquefied petroleum gas (LPG), and gas are viable substitutes for electricity in heating and cooking when available. Households that have few alternatives to heating with electricity have the greatest difficulty in shifting their energy consumption to less expensive fuels, making them more vulnerable to electricity tariff increases.

There are also sharp differences in electricity consumption between urban and rural areas. Poor rural households generally spend a lower share of income on electricity than poor urban households (figure 3.3), perhaps because they have greater access to inexpensive substitutes such as wood and coal, which can be used instead of electricity for heating, but which can have environmental and social costs. In addition, as the electricity supply deteriorated due to lack of investment and maintenance in the 1990s, rural areas may have been disproportionately affected by blackouts, contributing to lower expenditures.

Service Quality and Availability

Even though official statistics and household surveys suggest that access to service is nearly universal, supply is often rationed because of deteriorations in service quality. Some countries experience frequent interruptions in electricity supply and fluctuations in voltage that destroy household appliances.[13] Supply shortages are likely to become more widespread unless investments are made in rehabilitation and maintenance of infrastructure.[14]

Figure 3.3. The Rural Poor Spend Less of Their Income on Electricity Than the Urban Poor Do (2000)

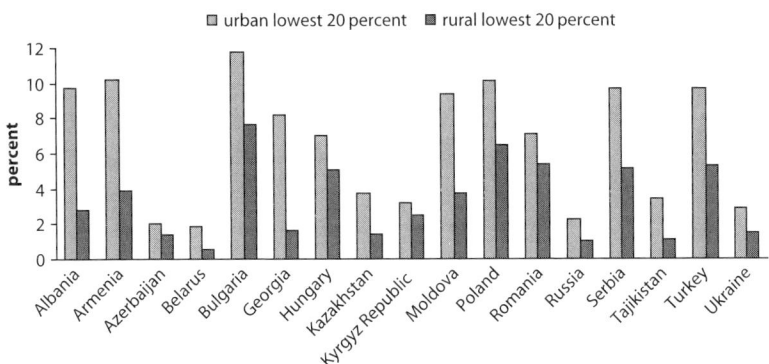

Source: Author's calculations based on household survey data from the World Bank ECA HBS database.
Note: Conditional on households reporting positive expenditures. Figures for Bulgaria and Tajikistan are for 2003.

Nonpayments

Nonpayment is one of the most vexing problems for electricity sector reform in ECA, and resolving it has been a key reform objective. But knowing who accumulates arrears is critical to understanding the welfare effects of reforms. If it is mainly the poor, affordability, not free-riding, is probably the cause of nonpayment. In fact, the poor are much more likely than the nonpoor to report zero electricity payments.[15] Nonpayments are positively correlated with expenditure ratios: the greater the electricity consumption as a percentage of total household expenditure, the more likely that the household does not pay its electricity bills (figure 3.4). This suggests that policies to raise collections and tariffs simultaneously will disproportionately affect the poor.

Other Energy Sources

Other Network Fuels: Gas and District Heating

Gas is an efficient alternative to electricity for heating and cooking, even at full import prices. While there may be additional costs associated with the technology required to use gas (metering and gas-fired appliances), the convenience and savings suggest that, given access, it is favored as a household fuel for heating and cooking. Back-of-the-envelope calculations confirm the rising use of natural gas. In Armenia, residential consumption of natural gas more than tripled from 1996 to 2001 (from 29,000 tons of oil equivalent to 90,000), while monthly electricity

Figure 3.4. Poor Households Are Less Likely to Pay Their Electricity Bills

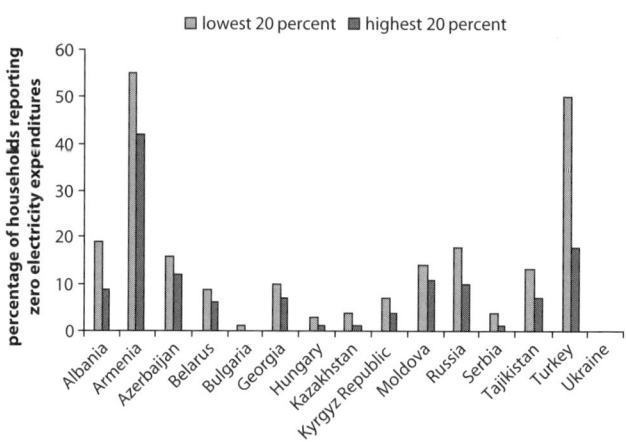

Source: Author's calculations based on household survey data from the World Bank ECA HBS database (2002). See annex 2 for more information.

consumption dropped from 187,000 tons of oil equivalent to 106,000.[16] For Georgia, the number of gas connections in the capital quadrupled from 2000 to 2003.[17]

The analysis for district heating and gas shows that while access levels are not as high as for electricity, expenditure patterns are similar. The lowest 20 percent of the population spends a higher share of income on district heating and gas than the top, and nonpayment rates are higher in the lowest 20 percent. In almost all countries, the highest 20 percent of the population has substantially higher access to gas than the bottom.

Non-network Fuels

If poor people are not using network energy, what are they using? A 2002 study found that in many cases the answer is traditional non-network energy.[18] Of non-network fuels, the cleanest, LPG, tends to be expensive, with coal and wood cheaper. Coal and wood use are consistently higher for the poor than the nonpoor (table 3.2).[19] Of seven country studies, in six the nonpoor are more likely to use LPG while the poor favor traditional non-network energy. Burning traditional fuels has environmental and social costs, including air pollution and deforestation, implying that reforms that raise prices of clean energy must take into account the size and economic implications of these costs.

Table 3.2. Urban Non-network Energy Use in ECA
(percent)

Country	Liquefied petroleum gas		Kerosene		Coal		Wood	
	Poor	Nonpoor	Poor	Nonpoor	Poor	Nonpoor	Poor	Nonpoor
Armenia, 1999	17	27	14	11	na	na	47	50
Croatia, 1997	44	45	3	7	1	1	51	26
Kyrgyz Republic, 1999	24	39	31	17	60	31	46	22
Latvia, 1997	37	28	na	na	<1	<1	1	2
Lithuania, 1998	na	na	na	na	<1	<1	1	2
Moldova, 1999	6	7	na	na	9	5	12	9
Tajikistan, 1999	na	na	<1	1	11	18	47	32

Source: Author's calculations from household survey data. Lampietti and Meyer 2002.
na = is not available from household survey.

Changes in Consumption across Income Groups[20]

The trade-off between income and energy expenditure can be assessed (preliminarily) by running a regression to estimate the expenditure elasticity of energy demand.[21] In doing so, however, it must be remembered that differences in rates of change in household spending across countries may be affected by differences in local policy and physical infrastructure.

Although there is variation across countries, the results show that, relative to income, poor people's overall energy expenditures are consistently more elastic than those of the nonpoor (figure 3.5). A 10 percent increase (decrease) in income results in an 8 percent increase (decrease) in energy expenditure for poor people and a 5 percent increase (decrease) for nonpoor people. (In other words, the elasticity of energy expenditure relative to income is 0.8 for the poor and 0.5 for the nonpoor.) In an environment of falling incomes in the late 1990s, the poor appeared to be cutting back on energy expenditures (as a percentage of income) faster than the nonpoor, probably by consuming less expensive traditional fuels.

This finding contrasts with the price elasticity of electricity expenditures, which is lower for the poor than for the nonpoor in the countries studied here. Why? Because in many places the poor already consume very low levels of electricity, and little scope remains for further reducing their consumption (chapter 9).

Conclusion

These findings tell us something about energy consumption patterns in ECA and how they have changed since the onset of transition and

Figure 3.5. Expenditure Elasticity of Energy Demand—Higher for the Poor than the Nonpoor

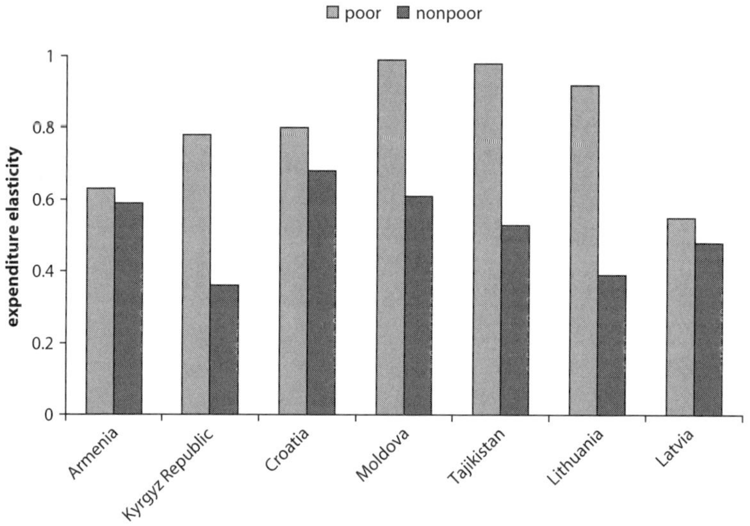

Source: Author's calculations from household survey data (Lampietti and Meyer 2002).

reform. For one of the main patterns, where poor households tend to choose non-network energy, there are two possible explanations. The first is that they do not have access to network energy such as gas, or have more restricted access to electricity as generation and distribution infrastructure deteriorate. In other regions this is often the case, but in ECA network energy use was high before the transition, indicating that network infrastructure was in place and almost all fuels were available in all countries.[22] The second explanation is that poor people choose non-network energy—wood, kerosene, or coal—because it is less expensive or because they do not have the resources to spend on appliances that enable them to use network energy, such as gas stoves.

The case studies that follow throw further light on how the poor are affected by reform and how to improve the design of reform to mitigate the impact on the poor and promote the use of clean energy.

Notes

1. World Bank (2001).

2. Unless otherwise stated, the household energy consumption and expenditure data in this chapter came from household budget survey (HBS) data from

both urban and rural areas in 17 of the 29 countries across the region for 2002. The HBS included questions on monthly per capita expenditure and access to and expenditure on electricity, gas, central heating, liquefied petroleum gas, total energy, and water. From this information it is possible to calculate electricity expenditure as a share of total expenditure (budget share), changes in budget shares before and after reform, and changes in consumer surplus. A full breakdown of all the household survey data is in annex 2.

3. See table 2.3.

4. The EU directive on power reforms includes liberalizing markets, unbundling utilities, and establishing regulated third-party access for the power network.

5. Poland, following the dissolution of the communist regime, embarked on an ambitious "economic transformation program" in 1990. But the Polish government has been more careful than Hungary to allow entry of foreign investors in the energy sector, deemed "strategic" by the government. For more information on Hungary's and Poland's EU requirements, see World Bank (1997b, 1999c).

6. Saavalainen and ten Berge (2003).

7. Esanov and others 2001. For suggestive evidence on this in other regions, particularly on the philosophical debate and empirical evidence on the inverse relationship between natural resource abundance and economic growth, see Sachs and Warner (1995).

8. Hellman (1998); Saavalainen and ten Berge (2003).

9. From 1993 to 2002, only 60 percent of International Monetary Fund (IMF) energy conditions were implemented (primarily relating to foreign energy debt and categorical privileges). Recognizing this, the IMF and the World Bank later reduced the number of conditions in all countries except Georgia (Saavalainen and ten Berge 2003).

10. The analysis in this section is based on the World Bank's ECA Household Budget Survey database for 2002.

11. Throughout the rest of the book, total expenditure is used as a proxy for income.

12. This finding holds across a wide number of countries, suggesting it is quite robust.

13. Markandya, Jayawardena, and Sharma (2001).

14. Cambridge Energy Research Associates estimates that half of Russia's generation capacity must be retired in the next 20 years as it reaches the end of its productive life, while more than the total installed generation capacity of France needs to be added. If these investments are not made, Russia is expected to suffer from nationwide electricity shortages in the near future (*The Economist* 2002).

15. Households may report zero payment for a variety of reasons, including poor service quality, billing cycles, and arrears.

16. Total residential consumption from the energy balance data in Armenia (Ministry of Energy). Converted to kilowatt hours (KWh) using the conversion factor of 1,000 KWh = 0.086 tons oil equivalent, this gives an increase in natural gas consumption from 337 million KWh in 1996 to 1,046 million KWh in 2001, and a reduction in electricity consumption from 2,174 million KWh in 1996 to 1,232 million KWh in 2001 (conversion factor from the World Energy Council) (Lampietti 2004).

17. Tbilgazi's customer base increased from 39,000 households in June 2000 to 164,000 households in January 2003 (Lampietti and others 2003).

18. Lampietti and Meyer (2002).

19. The exception is Tajikistan, where coal is heavily subsidized for everyone.

20. The analysis here, based on data from seven countries, was originally presented in Lampietti and Meyer (2002).

21. This follows the methodology in Subramanian and Deaton's (1996) study of the demand for food calories. Log energy expenditure is regressed onto log total household expenditure (a proxy for income).

22. See annex 5.

CHAPTER 4

Raising Prices in Armenia—What Happens to the Poor?

A pivotal moment in Armenia's electricity sector reform was a tariff increase in January 1999, several years into the reform program and after the height of Armenia's energy crisis. The increase was the most radical to date: it was the biggest, and it was a shift from an increasing block tariff to a single rate, removing the subsidy regime.

The move provoked an energetic debate among those working on reform at the World Bank, which had strongly encouraged Armenia's government to make the change. Those in favor of the increase argued that the current system, where the first 100 kilowatt hours (KWh) of electricity consumed was highly subsidized for all households, needlessly benefited the nonpoor and encouraged inefficient usage. Opponents of the increase argued that electricity expenditures constituted a higher percentage of the incomes of the poor, who would therefore be disproportionately affected by the price increase. Though the reform included a restructuring of the social benefits system, replacing the old tariff-based subsidy with a direct transfer, opponents were not convinced that this would be effectively targeted to the poor.

Adding significance to the debate were more general concerns about the effects of reform on the poor across the region.

Before the Price Hike

In the late 1980s and early 1990s, Armenia's economy suffered a catastrophic earthquake, the breakup of the Soviet Union, protracted conflict, and the closure of borders with Azerbaijan and Turkey Political and economic isolation—landlocked and entirely dependent on imported oil and gas—compounded the effects of rising energy prices. The cost of supplying electricity and central heating skyrocketed, while residential electricity prices remained very low. Unable to cover internal maintenance costs and crippled by the shutdown of the nuclear power plant and weekly interruptions in natural gas supply, by 1992 electricity utilities were on the verge of collapse.

Residential consumers bore the brunt of the utility crisis. From 1992 to 1995, most of the population received only two to four hours of electricity per day, and central heating and natural gas supplies were virtually terminated. The economy also suffered as public infrastructure and the industrial sector were hit by shortages. Consumers stopped paying their utility bills, and in 1994, payment for electricity fell to only 10 percent of billings, further threatening the sector's sustainability. With district heating also gone, residents of the capital Yerevan burned trees, telephone poles, and books to get through the winter, and deforestation for fuel wood took place on a devastating scale.

With an economic reform program in 1995, the economy began to stabilize, and starting in 1995–96, the Armenian government embarked on reforms to put the energy sector back on its feet. Armenia soon made progress in restructuring and regulating the energy sector, raising tariffs, improving payment discipline, and making the electricity supply more reliable. The result was a dramatic improvement in the supply of electricity; by 1999 most households were again receiving service 24 hours a day, and outages were shorter and less frequent.

Increasing cost recovery by utilities became a cornerstone of the government's economic reform program. Until 1999, Armenia had an increasing block tariff structure. The first 100 KWh of electricity consumed cost households dram 15 per KWh; the second block, 100–250 KWh, cost dram 20 per KWh; and the third block, above 250 KWh, cost dram 25 per KWh. The government's strategy—under pressure from the World Bank and the IMF, which were financing power sector rehabilitation—was to couple tariff increases with generalized social transfers targeted at low-income households.

On January 1, 1999, the Energy Regulatory Commission eliminated the increasing block tariff in favor of a single uniform tariff of dram 25 ($0.048) per KWh. The change in tariff structure led to a sizable increase in electricity prices, and to soften the impact of this increase, the poorest households were compensated with a direct cash payment through the social protection system.

Higher utility tariffs were already meeting political resistance, and in late 1998 and early 1999 the government was concerned about the impact of cost-recovery efforts on consumers, particularly the poor. The economy appeared to be on the path of sustainable growth, but transition, economic reform, natural disaster, and war had taken a heavy toll on the living standards of Armenians. Real wages were still only 12 percent of 1990 levels, and increases were outpaced by real prices of electricity and other utilities combined with a substantial increase in collection rates (figure 4.1). Further increases in tariffs and collection rates, unless effectively mitigated, would only add to the difficulties facing Armenians.

This study looks at who was more affected by tariff increases and the removal of tariff-based transfers (the poor or the nonpoor), how the effects show up, and whether transfers from the government's social benefit system cover the increase in the average tariff. By analyzing these questions, the study can help the government devise a socially equitable and politically feasible strategy for reform (box 4.1).

Figure 4.1. Electricity Price Increases Outpaced Real Wages

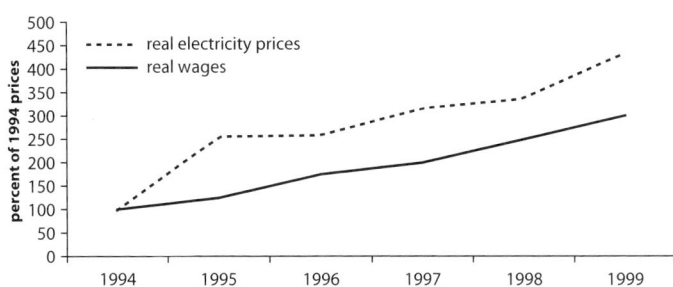

Source: Authors' estimates.

Box 4.1

Data for the Analysis—Armenia

Qualitative analysis was conducted through interviews and focus group discussions across the country to understand how people viewed reform and how they were dealing with it through substitutes and other coping mechanisms. This informed the quantitative data collection, which consisted of a survey of 2,010 randomly selected households from different parts of the country where people had access to gas, wood, and electricity, and was conducted in December 1999 and January 2000.

The household survey data were merged with Armenergo billing, payment, and consumption records from January 1, 1998 to December 31, 2000, to build a picture of household energy consumption patterns in response to changes in energy supply and increased tariffs. A subset of 1,514 households with complete utility records from March 1998 through December 1999 was used to analyze the impact of the 1999 tariff increase.[a] Unless otherwise noted, the discussion here compares household consumption and payment behavior between 1998 and 1999 using March 1998–November 1999 billing, payment, and consumption data.[b]

Note: a. In the analysis here, about 33 percent of rural and urban households are poor. The study draws poverty lines so that rural households with per capita expenditures of less than dram 4,100 and urban households with less than dram 6,700 are defined as poor. These are relative poverty lines defining the poor as the lower one-third of households in the per capita expenditure distribution and are generally consistent with the 1996 line. The proportion of poor and nonpoor households in the subset used to analyze the impact of the tariff increase is equal to the proportion in the overall sample, 485 poor and 1,029 nonpoor households. Unfortunately, complete billing records for households in Yerevan are available from the utility only for months after March 1998.
b. December 1999 billing and consumption data are excluded because payment information is not available for that month (January 2000 payments).

Residential Energy Consumption in Armenia

Uses of Energy

The surveys found that households consume energy for lighting, heating, water heating, and cooking. For lighting, 100 percent of households relied on electricity. For heating and cooking they consumed wood, electricity, central gas, liquefied petroleum gas (LPG), and, much less commonly, dung, kerosene, and diesel.

Large amounts of energy were consumed for heating, particularly in the colder winter months. Fifty-three percent of households used wood, and 17 percent electricity (table 4.1). Other important heating sources

Table 4.1. Wood and Electricity Make Up the Bulk of Household Winter Energy Expenditures (percent of household expenditure)

Primary source	Heating		Heating water		Cooking	
	Poor	Nonpoor	Poor	Nonpoor	Poor	Nonpoor
Electricity	10	20	31	37	28	29
Central heat	7	9	—	—	—	—
Central gas	8	11	11	13	13	16
LPG	<1	<1	1	1	8	13
Wood	59	50	47	43	41	35
Dung	6	5	6	5	6	4
Other	9	5	4	2	5	2
Total	100	100	100	100	100	100

Source: 1999–2000 survey. See box 4.1.

Note: This is household expenditure, not per capita expenditure. Poor households are typically larger than nonpoor households, so per capita expenditure gives different results.

included natural gas and central heat. Rural households depended more on wood for heating than urban households. In urban areas, the poor depended significantly more on wood for heating (56 percent) and less on electricity (14 percent) than the nonpoor (42 percent and 29 percent).

For heating water, 44 percent of households used wood, 35 percent electricity, and 14 percent natural gas. Again, the urban poor depended significantly more on wood and significantly less on electricity than the urban nonpoor for both heating water and cooking. Natural gas was favored for cooking by the nonpoor in both rural and urban areas.

Energy Consumption and Expenditure

The poor consumed 20–30 percent less of each energy type than the nonpoor. Median annual household consumption of electricity was 1,275 KWh, LPG 60 kilograms, wood 5 cubic meters, and dung 5 cubic meters.[1] Rural households consumed less electricity and natural gas and more wood and dung than urban households.

Though the poor consumed less, the burden of energy expenditures was particularly large for them. Poor households devoted close to 30 percent of their monthly expenditure to energy, compared with 18 percent for the nonpoor (table 4.2).[2] Electricity made up the bulk of energy expenditures for all households. The burden of tariff increases appeared to be highest among the urban poor, with 16 percent of their total monthly expenditures going to electricity alone. The rural poor spent equivalent amounts on wood and electricity. In western countries, including North America and Western Europe, electricity expenditures typically range from 3 percent to 7 percent of total income.[3]

Table 4.2. The Burden Is Higher for the Poor
(percent of monthly expenditure)

| | Rural | | Urban | |
	Poor	Nonpoor	Poor	Nonpoor
Electricity	13	7	16	9
Natural gas	6	4	3	2
Wood	13	8	5	2
Other	1	1	2	1
Total	34	21	27	16

Source: 1999–2000 survey. See box 4.1.
Note: Households reporting positive expenditures for at least one of the three energy sources in the table.

Remember that expenditures on wood and dung might not fully reflect consumption, particularly for the poor, since households often collected part or all the wood they use for heating and cooking rather than buy it. Sixteen percent of households cut their own wood, with this activity concentrated in the densely forested province of Lory Marz. The poor spent approximately 20 days a year on this activity, the nonpoor 12 days.

Improvements in Electricity Supply

The reliability of electricity supply increased steadily after 1994, when service was available for only a few hours a day. In 1996, slightly less than 90 percent of households reported expenditures on electricity. By 1999, 98 percent of households reported having electricity and paying for it, with the remaining 2 percent reporting that their service was cut off due to nonpayment.

In addition to better coverage, electricity service became more reliable. In focus groups and individual interviews, respondents indicated that electricity was available 24 hours a day in all locations and that there were no electricity failures. Seventy percent of respondents reported no break in service in the preceding year. Of the 30 percent who did have a break, the median duration was 3 days and the mean 14 days. These figures were slightly higher for the poor (4 days and 17 days) than for the nonpoor (3 days and 12 days). The quality of service was rated as average or better than average 95 percent of the time, though 22 percent of respondents reported having lost an appliance in the last year as a result of a surge in electricity.

Respondents were satisfied with the electricity service maintenance. Eighty-six percent believed that a utility employee performs maintenance activities. But the qualitative consumer satisfaction survey indicated that respondents were not sure whom they should contact if they had problems with their electricity service. Friends or relatives often made informal repairs in the household, while a higher percentage of the poor reported making repairs themselves (13 percent) than the nonpoor (7 percent).

How Households Cope with Increasing Collections

Armenia made considerable progress in improving electricity collection rates after 1994. Electricity is metered in all households, with meters read at least once a month.[4] But with energy expenditures accounting for 15–30 percent of per capita income, many households have a very difficult time paying bills. Focus group discussions suggested that one coping mechanism was to pay only a fraction of the bill, maintaining service while accumulating arrears. Another coping mechanism was to monitor consumption closely and then impose austerity measures when the budget is reached.

Most households paid only part of their electricity bill each month, contributing to a rapid increase in arrears. When questioned directly, households reported that they paid to avoid having their service cut off.

A common explanation from households for nonpayment was that the government owes them salary arrears, pension arrears, or the savings that disappeared from their bank accounts at the end of 1993. Public sector employees often stated that until wages are increased they would not pay their utility bills.

How did people reduce consumption of electricity in the face of rising prices? Ninety percent of the poor and 86 percent of the nonpoor said they were always careful to turn out the lights when leaving a room. Seventy-three percent of the poor and 61 percent of the nonpoor said they always made an effort to wear more clothing to reduce the amount of electrical heating they consume. In response to higher prices, people became more economical with their consumption. The question is: At what point does this change in behavior become a burden that reduces welfare?

Use of Substitutes

Eighty percent of all households and 95 percent of the rural poor reported that they substituted for electricity in response to rising prices, primarily for heating and cooking. As electricity consumption dropped, reported consumption of wood and natural gas increased. More than 60 percent reported that the primary substitute was wood and about 24 percent gas (table 4.3). The stated increased reliance on wood was particularly acute

Table 4.3. How Are Households Reducing Reliance on Energy?
(percent)

	Rural		Urban	
	Poor	Nonpoor	Poor	Nonpoor
Wood	63	68	68	57
Natural gas	14	16	17	35
Kerosene	1	1	7	6
Dung	18	14	1	1
Other	3	0	7	7
Total	100	100	100	100
Number of households	93	204	181	316

Source: 1999–2000 survey.

among the urban poor. When asked if they made an effort to reduce their reliance on electricity over the previous 12 months, about 65 percent of the poor and 54 percent of the nonpoor said they had, with the effort highest among the rural poor (71 percent).

Data on the price of substitutes were also collected as part of the survey (table 4.4). While electricity and natural gas prices were constant at dram 25 per KWh and dram 51 per cubic meter, there was some geographic variation in the prices of LPG (between 300 and 400 drams per kilogram, highest in urban Lory Marz and lowest in rural Ararat Marz) and wood (ranging from a low of dram 4,000 per cubic meter in densely forested Lory Marz to a high of dram 8,000 per cubic meter in Yerevan).

While the inefficient practice of heating with electricity had been reduced, this had to be balanced against potential environmental problems associated with increased wood consumption, such as deforestation and increased indoor air pollution.

Table 4.4. Prices of LPG and Wood, December 1999
(drams)

	LPG (drams per kg)		Wood (drams per m³)	
Marz (province)	Urban	Rural	Urban	Rural
Yerevan	322	—	8,104	—
Gegarkunik	332	341	6,246	6,721
Lory	370	334	4,288	4,092
Ararat	329	318	7,971	7,816
Shirak	328	350	6,568	7,409

Source: 1999–2000 survey.
— is not applicable.

Increases in wood consumption appeared closely correlated with the price of wood, with the highest stated increases in wood consumption in Lory Marz, and the lowest increase in Yerevan. Again, this suggested that the burden of rising electricity prices was likely to be highest for poor urban households, who faced the highest priced substitutes for electricity. Although nonpoor households consumed more electricity, as a percentage of household monthly expenditure the poor were disproportionately affected by the price increase.

Attitudes to Reform

The qualitative data yielded interesting information about public opinion and awareness of the electricity sector, which backed up the government's ambivalence toward tariff reform. Although affected by power outages and worried about the safety of the nuclear power plant, focus group participants worried that utility sector improvements would be too expensive to implement and were concerned about how utility reforms would affect the poorest segments of the population. Corruption was perceived as a major obstacle, suggesting that the government had limited credibility and could not rely on public support for continuing policy changes.

Who Suffered Most: The Impact of Reform

The analysis provides a comprehensive snapshot of household energy consumption in Armenia. The discussion now turns to look more closely at the impact of the January 1999 change—from a tariff-based subsidy, in the form of an increasing block tariff, to a uniform price—on household electricity consumption and payment behavior.

Magnitude of the Tariff Increase

How much of an increase did the shift to a single tariff represent? In 1998, before tariff restructuring, the government reported the effective average household tariff of dram 19.2 per KWh. This calculation was based on aggregate utility data, dividing total bills by total consumption for all households in 1998. Based on this average, the hike in the tariff to dram 25 per KWh represented an increase of 30 percent.[5] But a more accurate measure of the effective average price facing individual households under the old tariff structure could be calculated only by using individual household consumption and billing records. Taking the average of the effective price facing each individual household during 1998 using available monthly electrical utility billing records for the survey sample produced an effective price of dram 17 per KWh. Based on this average, the 1999 tariff change could be seen as a 47 percent increase in price—over 50 percent more than originally thought. (Clearly, price response prediction can be much improved through better methods, data reporting, and sector statistics.)

Overall Impact of the Price Increase

Following the tariff increase, total household electricity consumption dropped 17 percent, from 2.2 million KWh in March–November 1998 to 1.8 million KWh during the same months of 1999 (table 4.5). Despite this drop in consumption, the new tariff resulted in a 16 percent increase in total billings. But utility revenues from the households increased only 6 percent, as their payments failed to keep pace with billings. Calculated collection rates, the ratio of total payments to total billings, fell 9 percentage points, from 97 percent in 1998 to 88 percent in 1999.

Both the collection rate and the change in the collection rate reported in the analysis here are higher than the 86 percent and 79 percent reported by the government for 1998 and 1999. There are three possible explanations. First, the analysis does not include data from the months with the highest incidence and level of arrears—January, February, and December—thus resulting in a potential overestimation of collection rates (for example, the collection rate for 1999 calculated for bills for January–November is only 85 percent). Second, the government's reported figures include technical and commercial losses as well as nonresidential consumption and payment data not included in household billing data used in this analysis. Third, the government's reported figures are based on national data, whereas the analysis here is based on a sample from the five marz, or administrative divisions. The sampled marz may have systematically higher collection rates than the rest of the country.

Table 4.5. Aggregate Impact of Electricity Tariff Change

Household[a]	*1998*	*1999*	Change between 1998 and 1999	
			Units	*Percent*
Consumption—million KWh	2.22	1.83	−0.38	−17
Billings—million drams	39.57	45.79	6.22	16
Payments—million drams	38.22	40.33	2.11	6
Collection rate—percent	97	88		−9[b]

Source: 1999–2000 survey.
a. For sample households only.
b. Percentage points.

Effects on the Poor and Nonpoor

Average household consumption by the nonpoor fell by 16 percent—from 178 KWh per month during March–November 1998 to 141 KWh per month during these months in 1999.[6] Poor households responded more strongly to the price change, lowering their consumption by an average of 20 percent from 152 KWh per month in 1998 to 121 KWh per month in 1999, enough for a refrigerator and a few lightbulbs. This suggests that consumption was more elastic among the poor, until the minimum basic consumption level was reached. Consumption declined significantly more among rural households (26 percent) than urban (13 percent), probably because rural households had greater access to substitutes.

Effect on Bill Amounts and Payments

For nonpoor households, although average consumption fell 16 percent, average monthly bills under the new tariff increased 17 percent—from dram 3,010 in 1998 to dram 3,520 in 1999. But these households increased their average monthly payments to the utility by only 7.5 percent—from dram 2,970 to dram 3,190.

Despite a 20 percent reduction in average consumption by the poor, their average bills increased by 13 percent—from dram 2,680 in 1998 to dram 3,020 in 1999. Average payments by the poor remained about the same at approximately dram 2,450 a month. The observation that average expenditures by poor households were more or less constant before and after the price change suggests that the poor could not or would not spend more than they currently do on electricity.

The gap between billings and payment—the arrears—increased significantly for both the poor and the nonpoor between 1998 and 1999. From March to November, total arrears increased fourfold—from dram 1.4 million in 1998 to dram 5.5 million in 1999 (figure 4.2). In 1998, the nonpoor accounted for less than a quarter of arrears even though they constituted two-thirds of the sample population. In 1999, arrears of the nonpoor grew dramatically, accounting for more than half of the total. Two factors contributed to this increase. First, the number of households not paying their bill in full each month increased; in 1998, on average, fewer than a quarter of the households did not clear their bills in a particular month. In 1999, this figure went up to more than one-third of the households. Second, the average monthly size of the unpaid balance per household increased by 13 percent.

Figure 4.2. Arrears Increased for the Poor and Nonpoor

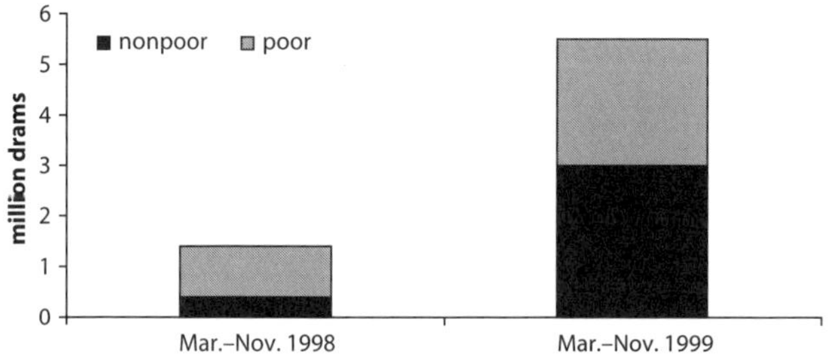

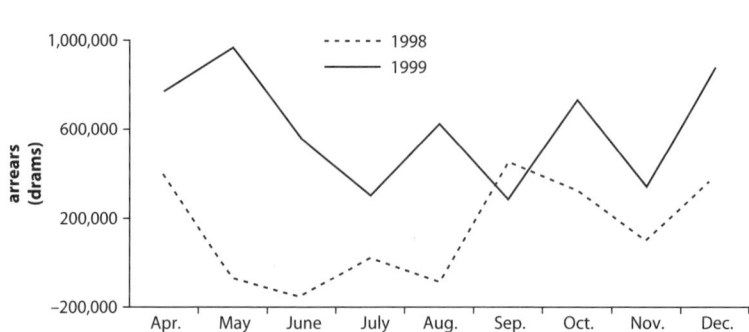

Source: 1999–2000 survey.

Arrears Levels for the Poor and Nonpoor

The percentage of households carrying arrears increased more among the poor than the nonpoor.[7] Among the nonpoor, the percentage of households carrying arrears increased from 22 percent to 37 percent, while that of poor households increased from 27 percent to 46 percent, 15 and 19 percentage points respectively. The increase in the size of arrears was larger among the poor than among the nonpoor, 15 and 10 percent respectively. That arrears were larger and increased more for the poor suggests that affordability, not free-riding, was the problem.

Monthly billing and payment trends in 1998 and 1999 suggest that households had the most difficulty paying bills in the winter when consumption was higher (figure 4.3). In 1998, before the tariff change, households paid off their winter arrears from May until August. In 1999, however, households were unable to pay off their winter arrears—they

Figure 4.3. Arrears to the Utility Went Up

Source: 1999–2000 survey.

tended to accrue additional arrears during the summer months, accumulating significant debt to the utility over the year.

How Effective Was the Cash Transfer?

The Armenian government took two actions to minimize the impact on the poor of the 1999 tariff increase. First, a newly designed family benefit, targeted at the 28 percent of households living below the poverty line, was introduced in 1999 as part of the government's reshaping of the family benefits system and to help alleviate the impact of the tariff increase. Second, an additional 9 percent of households not eligible for the family benefit, but expected to have difficulty paying their electricity bills, received a smaller sum, dram 1,450 per month, to assist with electricity payments.

The study data confirm that about 28 percent of households received the family benefit, with the mean monthly amount received at drams 9,480 per household, or dram 2,500 per capita. Those data also show that about 8 percent of households received a special cash benefit or other electricity privilege of some kind in 1999.[8] Almost all of these latter households reported receiving dram 1,400 or 1,450 per month, though they only received it an average of six times during the year, making an average of dram 9,470 per household per year. This is less than the targeted dram 17,400 a year with 12 months of payments, but still represents a significant cash transfer—almost five months of the average 1999 electricity bill for these households.

Targeting Effectiveness of the Cash Transfers

The study team did not have access to information on the formula used to determine which households were targeted to receive the cash transfers, so it was not possible to determine the success of targeting and whether the family benefit was indeed going to the poorest households.[9] But it was possible to examine whether poor households and households regularly consuming in the lower blocks of the 1998 tariff structure reported receiving the transfer. Poor households were more than twice as likely to receive the cash transfer as nonpoor households, and households regularly consuming electricity in the first two blocks of the 1998 tariff were significantly more likely to receive the cash transfers. But only 55 percent of the poor received the cash transfer, meaning that 45 percent of poor households were faced with a 47 percent increase in their electricity tariffs and no mitigating cash transfers.

Effectiveness of Transfers in Softening the Impact

The cash transfers did help soften the impact for those receiving them. Households receiving the cash transfer cut their consumption after the price increase by about 20 percent—similar to poor households overall. And their average bills rose by 15 percent, again similar to poor households overall. But unlike the other poor households—whose average payments were unchanged between 1998 and 1999, and which therefore accumulated even more arrears—households receiving cash transfers increased average monthly payments to the utility by 4 percent. It is difficult to determine whether the cash transfers offset the adverse impact of the tariff increase. However, it may well have prevented an even greater drop in consumption and an increase in arrears among the recipients. It is also possible that these cash transfers may work even better if targeted households receive them every month.

Conclusion

In creating a picture of the household response to Armenia's 1999 electricity tariff change, this study reveals that restructuring the tariff to a less socially regressive single rate had a disproportionately negative impact on the poor. The burden of energy expenditures, the bulk of them for electricity, was large for most households and particularly for the poor. Relative to the nonpoor, the poor cut consumption more, the percentage of poor households with arrears was higher, and the average size of their arrears increased more. The tariff increase was 50 percent greater than originally conceived when the increase and mitigating transfers were formulated, so the impact was underestimated—spotlighting the need for careful calculation and accurate price response prediction in forecasting and mitigating the impact of reform.

Though the poor were meant to be compensated with cash transfers, both the targeting and the timeliness of the transfers needed to be improved. Opponents of the increasing block tariff were correct in arguing that it benefited 100 percent of consumers when only 33 percent were classified as poor. But with the new tariff structure, only 55 percent of the poor actually received the social benefit transfer, leaving almost half of them uncompensated for the 47 percent tariff increase. With limited access to low-cost substitutes, a further increase in tariffs and collection rates would lead to the greatest hardship for the urban poor, who spend 16 percent of monthly cash expenditures on electricity and have the least access to wood.

As electricity consumption dropped, reported consumption of natural gas and wood during the period increased. Use of wood is associated with such environmental problems as deforestation and indoor air pollution. These serious concerns indicate that governments must be prepared with policies to identify economically efficient and sustainable actions to meet basic heating needs as tariffs are increased and develop a long-term national heating strategy.

Although the tariff increase was aimed at creating a more sustainable sector, the utility revenue increase of about 6 percent from sampled households was less than expected, thanks to falling consumption and a simultaneous increase in arrears. This suggests that the benefits of the reform program did not materialize as quickly or easily as intended, and that tariff increases must be accompanied by moves to encourage greater payment. That the fall in collections coincided with political turbulence in 1999 points to the importance of consistent government support and political stability in successful reform, issues that come up in other case studies.

This study provides valuable insights on the impact of reform in the short term; since it was conducted, reform has had more time to take hold. Armenia has continued to reform and reorganize the electricity sector, transforming it from imminent collapse in the mid 1990s to one of the region's success stories. The government made abortive attempts to sell the loss-making distribution network to foreign strategic investors before selling a controlling stake in 2002 to Midland Resources Holding, which managed to significantly improve collection rates. By 2004, collections had reached almost 100 percent.[10] Since 2002, Armenia has had a national heating strategy. Use of gas for heating has increased, while use of wood has declined. The economy has continued to grow and poverty has declined significantly since 1999. Meanwhile the social protection system has become better targeted since 1999, and efforts to improve it further continue.[11]

This study does much to illustrate that understanding the impact of tariff changes on the poor has been imperfect at best. It provides empirical evidence to substantiate concerns about the speed of reform, suggesting that though reform was indisputably necessary, a fuller understanding of its effects could inform a more effective mitigating strategy. The study looks only at the short-term effects on the poor, raising the obvious question of how the story changes when the effects of reform are studied over a longer period. It also raises other questions that could not be answered with the data available. How can cost recovery be improved while maintaining a balance with social protection? If affordability is an issue, as it appeared to be in Armenia, how can tariffs be increased at the same time as collections? And what difference, if any, does it make if a private operator, more explicitly motivated to improve revenues, enters earlier in the-process? To address these questions, the discussion turns to Georgia.

Notes

This chapter is based on Lampietti, Kolb, Gulyani, and Avenesyan 2001.

1. Conditional on positive consumption of a given type of energy. Wood consumption is substantially lower than reported in the qualitative portion of the survey, where households said they consumed 20–30 cubic meters a year.

2. This result must be treated with caution because the mean is influenced by a number of high consumption and low expenditure values.

3. Kaiser (1999). Kaiser also reports that in 1997, an average of 9–10 percent of household income was spent on electricity during the winter and summer in Armenia. This is broadly consistent with data for Armenia from 1996, which suggest that the cost of electricity exceeded 10 percent of an urban household's expenditures for the average very poor family and less than 3 percent for a nonpoor family (World Bank 1996a). The poverty lines are set differently and expenditures are measured differently, so they are not directly comparable.

4. In 85 percent of households, meters are read once a month and, in the remaining 15 percent, an average of twice a month. The poor are much more likely to have their meter read twice a month (22 percent) than the nonpoor (12 percent).

5. Performing this calculation with the utility billing record produces the same figure of dram 19.2 per KWh. According to the World Bank's Project Appraisal Document for the Electricity Transmission and Distribution Project in Support of the First Phase of the Power Sector Restructuring and Development Program (February 8, 1999), the increase was from an average household tariff of dram 19.8 per KWh (p. 16).

6. The drop in consumption does not appear to be caused by climatic variations because temperatures during the major heating months in the period of analysis were actually lower in 1999 than in 1998.

7. A household is considered in arrears if the difference between the payment and the bill is greater than 5 percent of the bill.

8. Five percent of households said they received the special cash benefit while 3 percent said they received a reduction in their bill or a voucher (usually for 600 KWh) to help defray their electrical bill. Only 10 percent of households receiving the special cash benefit, 9 of the 2,010 sampled, claimed to be receiving both the family benefit and the special cash benefit.

9. An important future step will be to analyze this information and compare it with this report's findings.

10. World Bank (2004e). In 2005, Midland Resources Holding sold the distribution company to RAO UES for a sizable profit.

11. World Bank Poverty Reduction Support Credit for Armenia, October 21, 2004.

CHAPTER 5

Nonpayment and Power—Georgia

The severity of Georgia's electricity crisis in the late 1990s—when even households in the capital were receiving fewer than six hours of electricity a day, and collections were almost nonexistent—created an impetus for relatively rapid reform and privatization. In 1998, in the first privatization of its kind in the former Soviet Union, an American distribution company purchased Telasi, the Tbilisi power distribution company. From the outset, its main challenge was to increase revenue to cost-recovery levels. Over the next five years, battling low collections, high theft levels, and diminishing political will to back reform, this task proved almost impossible. Amid mounting concerns that low collections were threatening the future of reform in Georgia and putting other strategic investors off investing in the region's energy sector, this study set out to identify the factors that were driving household electricity consumption and payment behavior. It was hoped that this analysis would highlight effective strategies to increase collection rates and improve cost recovery, while maintaining a balance with the social and environmental effects of reform.

Deep Declines—Then High Expectations

Beset with civil war and the loss of central transfers, Georgia's economic decline following independence was among the deepest in the former Soviet Union, with gross domestic product (GDP) falling by 70 percent from 1990 to 1994. With the end of civil war in 1994, the government started a program of economic reform, and in the late 1990s the economy stabilized. But recovery was slow to translate into better living standards for Georgia's 5.4 million people, and poverty remains widespread.[1]

In common with citizens in other energy-poor republics in the region, Georgians faced higher costs and deteriorating service for household utilities, particularly energy. Georgia's dependency on energy imports and high international prices for fuel were exacerbated by supply and generation disruptions from political turmoil. Utilities accumulated large payment arrears, and energy supplies contracted dramatically. By 1997, electricity supply was 40 percent of peak levels and strictly rationed, and district heating was no longer in service. Georgia was experiencing an energy crisis.

Starting in 1996, with the support of the World Bank and other donors, the government of Georgia undertook a seemingly model program of utility sector reform (figure 5.1).[2] Sakenergo, the vertically integrated electricity enterprise, was split into several generation enterprises and separate transmission and dispatch companies. Distribution was divided into regional companies.[3] And Telasi, the electricity distribution company serving Tbilisi, was sold to the American power generation and distribution company, AES Corporation.[4]

Figure 5.1. Milestones of Power Sector Reform in Georgia

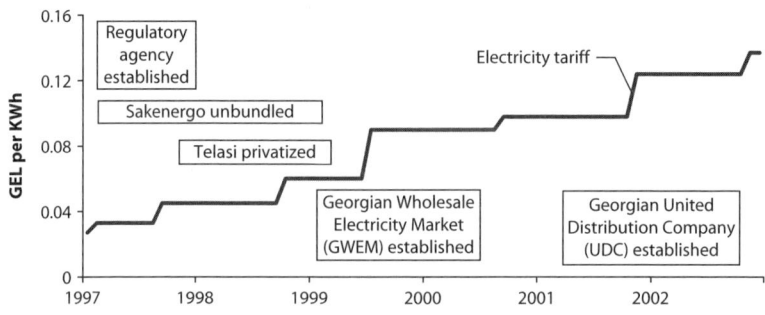

Source: Georgia National Energy Regulatory Commission (Annual Reports) and personal interviews.

Reform, particularly the high-profile entrance of an American company, brought high expectations for improving the situation. When AES took over Telasi in December 1998, only about 15–30 percent of the sector's generation capacity was operational. Households were receiving only four to six hours a day of electricity in Tbilisi, and three to four hours a day elsewhere.[5] Investment in maintenance and repair of electricity infrastructure was impeded by a lack of capital, as a combination of subsidized tariffs, nonpayment of bills, and thefts of electricity contributed to low cost recovery. To turn this situation around and increase collections, AES Telasi adopted a strategy of investing significant resources in remetering households in Tbilisi and cutting off dangerous illegal connections. As an incentive for payment, it promised 24-hour supply to households who paid their electricity bills.

But falling incomes and a prevailing practice of nonpayment—with high theft levels, routine sabotage or destruction of meters, and protests against increasing collections—proved to be major obstacles to improving cost recovery. On top of this, corrupt and inefficient elements within the supply system were undermining incentives to pay by diverting electricity to some nonpaying public sector customers while depriving AES Telasi of sufficient power for its paying customers, particularly during periods of high demand in winter.[6]

In 2002, when this study was commissioned, reform had stalled. Dissatisfaction with higher tariffs and greater enforcement was expressed through resentment at the presence of a western player in the electricity sector. In response to sustained operating losses, AES repeatedly threatened to withdraw from Georgia. Donors, losing confidence in the sustainability of the reform program and the probability of further reform, were assessing options for moving forward. Regional geopolitics, and the prospect of expanding Russian control over energy markets in the former Soviet Union, meant that the prospect of AES being replaced by a Russian operator was not viewed with total equanimity by western donor governments. And the experience of AES seemed to be part of a worrying trend across the region, as private interest in utility investments ebbed with changes in the world economy and a bursting of the privatization bubble. The Armenia study prompted hopes that analyzing what had and had not worked—and taking a closer look at the dynamics of utility reform—might help resolve some of the obstacles and mark out an approach for moving forward. This analysis would be important for sustaining reform in Tbilisi and for the future of reform in other parts of Georgia still facing severe shortages in energy supply.

This study was conducted six years after reform began, so it could use a richer data set than the Armenia study and analyze trends over a longer period. This provides a more nuanced and comprehensive picture of effects, such as changes in consumption, coping mechanisms, fuel substitution, access to alternative fuels, and factors determining household welfare. The study looks at a wider range of indicators and effects, including the impact of social transfers on the government budget. It also looks at some key questions on reform design in countries with low collections, analyzing what contributed to increased collection rates and suggesting how utilities can increase payments. And it assesses the targeting success of the direct transfers to mitigate the impact of reform and shows how the household and utility data can be used to improve targeting of the transfer.

Box 5.1

Data for the Analysis—Georgia

Since the quantitative data already existed, they were analyzed first, followed by qualitative analysis to better understand the findings. The data came from three sources: the household budget survey (HBS), conducted quarterly since 1996 by the Georgia State Department of Statistics;[a] the Multi-Sector National Survey of Households in Georgia 2002,[b] carried out by Save the Children (STC) in February 2002; and the electricity consumption, billing, and payment data from AES Telasi for those households in Tbilisi that were included in the HBS from 2000 to 2002. Since the data concentrated on Tbilisi, and because the situation in rural areas is rather different from that in the capital, the focus of the study is, for the most part, limited to Tbilisi.

Merging the HBS data (based on households' self-reported electricity payments) and AES Telasi data sets (based on household payments recorded in the customer's billing and payment records) revealed important discrepancies in reported electricity payments. A comparison of the corresponding data (for the same household in the same month) revealed that payments reported in the HBS were consistently higher than those recorded by Telasi in 2000 and 2001 (box figure B5.1). This difference could have been caused by corruption—for example, households paying more to the meter readers than is transferred to the utility—or recall error.[c] Despite these differences, the data sets provide a sound basis for the analysis because both follow the same increasing trend in payments and the difference between the two narrows over time.

(Continued)

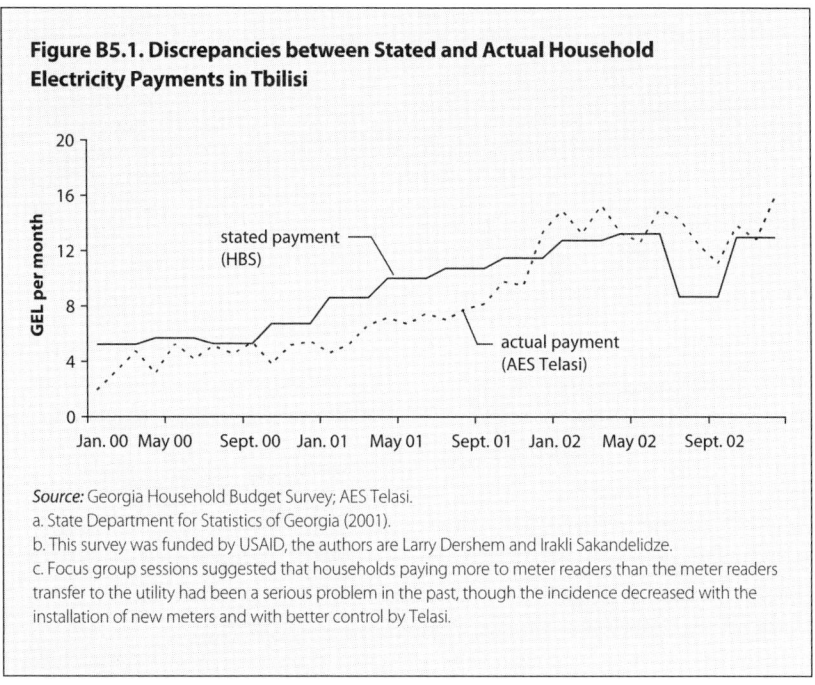

Figure B5.1. Discrepancies between Stated and Actual Household Electricity Payments in Tbilisi

Source: Georgia Household Budget Survey; AES Telasi.
a. State Department for Statistics of Georgia (2001).
b. This survey was funded by USAID, the authors are Larry Dershem and Irakli Sakandelidze.
c. Focus group sessions suggested that households paying more to meter readers than the meter readers transfer to the utility had been a serious problem in the past, though the incidence decreased with the installation of new meters and with better control by Telasi.

Residential Energy Consumption in Georgia

Availability of Energy

While 98 percent of households remained connected to the electricity network, supply was still failing to meet demand even after almost a decade of reforms in 2003.[7] In February and March 2001, less than half of total demand in Tbilisi was supplied, though service in Tbilisi has generally improved over the past few years. Outside Tbilisi, supply constraints are severe and persistent, with households in 2002 receiving 4.5 to 17 hours of electricity a day, depending on location.[8]

For natural gas, the number of connections increased in Tbilisi, particularly in 2001 and 2002.[9] Outside the capital, however, the number of connected households fell, possibly because of limited or nonexistent service and an extremely dilapidated gas infrastructure.[10] District heating disappeared in the late 1990s.

Changes in Relative Energy Prices

Electricity tariffs have more than doubled since 1997 in nominal terms.[11] This has a dramatic effect on the relative prices of other fuels, affecting household energy consumption choices.[12]

Until recently, residential natural gas tariffs remained fairly constant at GEL 0.27 per cubic meter in Tbilisi and GEL 0.30 per cubic meter in other cities, and even at full import prices gas was much less expensive than all other fuels. This, and the convenience of using it, suggests that it was the household fuel of choice for those with access. But households wishing to connect (or reconnect) to the gas network in Tbilisi had to pay GEL 215 ($100 in 2002), either up front or over time to cover the cost of a meter, or be billed GEL 6.50 per person a month.[13] Some participants in focus group sessions said that this high up-front connection cost, along with the need to invest in new gas-fired appliances, was a barrier to installing gas in their homes.

Clean network fuels—electricity and gas—had lower prices than non-network fuels, liquefied petroleum gas (LPG) and kerosene (figure 5.2). The latter actually became much more expensive over the period from January 1997 to January 2002.[14] Kerosene, an inferior fuel, is by far the most expensive fuel and therefore the least likely choice.

An important omission in this comparison of energy prices is fuel wood. It is commonly used, but there are no reliable data on price changes over time. Estimating wood prices is complicated by regional differences in availability, and thus price, and by the fact that households can either collect wood or buy it whole or split. The HBS collects information only from households that have purchased wood. This results in underestimation of consumption, since the Save the Children survey found that, depending on the region, anywhere between 5 percent and 75 percent of households cut their own wood. And there are important

Figure 5.2. Clean Network Fuels Cheaper Than Non-network Fuels
(effective energy prices, GEL per million Btu [British thermal unit])

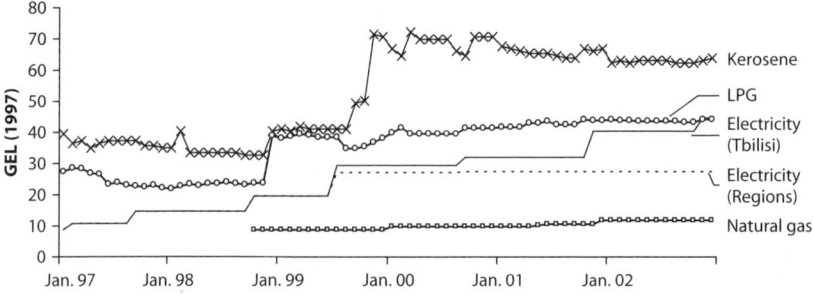

Source: Georgia State Department for Statistics.

differences in access for households that cut wood themselves—and thus in the time-related opportunity costs of collection.

The data give some idea of the relative price of wood. Survey research indicates that in the winter of 2002, wood prices were on the order of GEL 22 per cubic meter. Assuming a typical conversion efficiency of 20 percent, the cost of wood energy would be GEL 15 per million Btu—less than for all other fuels except natural gas. For poor households not on the gas network, that makes wood the fuel of choice for cooking and heating.

Effect of Reform on Energy Consumption

In Tbilisi, the highest 20 percent of households' energy consumption initially dropped, but eventually recovered to preform levels, at about 200 million Btu per quarter. The lowest 20 percent of households maintained the same consumption, at about 55 million Btu per quarter (figure 5.3). Fairly stable energy expenditure shares and consumption levels suggested that households in Tbilisi, in response to tariff increases, appeared to be replacing electricity with less expensive fuels.

Breaking down total expenditures into its parts reveals just this pattern. In Tbilisi, households increased the share of electricity in total energy from 45 percent to 51 percent from 1996 to 2002, and from 3 percent to 7 percent of total expenditure. The share of kerosene dropped, and those of LPG and purchased fuel wood stayed constant. The share of gas increased from 2 percent to 20 percent of energy expenditure, with the greatest increases at the end of 1999.

Figure 5.3. Energy Consumption Remained Low for the Lowest 20 Percent in Tbilisi

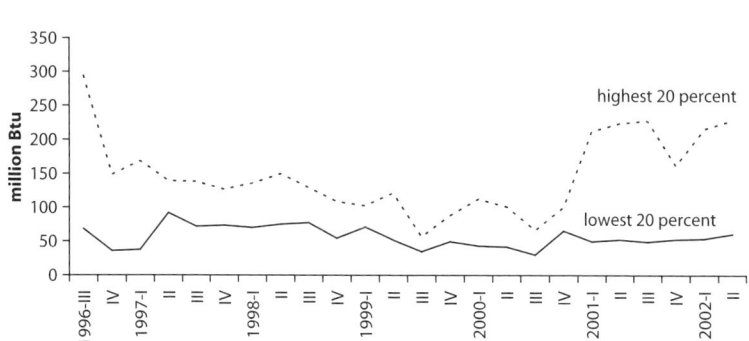

Source: Georgia Household Budget Survey. (See box 5.1).

Outside Tbilisi, energy consumption fell significantly after 1997. There was some stabilization in mid-1999, but the highest 20 percent of households now consume one-third as much energy (in effective Btus) as they did in 1997, and the lowest 20 percent of households about half as much. Unlike in Tbilisi, substitutes are not widely available. Fuel wood and kerosene remain significant in energy expenditures—and since kerosene is expensive, its use may be a response to inadequate electricity supply. Similarly, the reduction in overall consumption can be attributed to budget constraints and the lack of opportunity to substitute lower cost fuels, such as electricity and natural gas, for kerosene and LPG. This pattern of consumption implies that an improvement in electricity and gas supplies at their current prices is likely to result in welfare gains for households outside Tbilisi.

Changing Household Energy Expenditures

Although energy prices increased, the average household share of expenditure spent on energy remained constant at 8 percent. But it increased most in cities: in Tbilisi from 6.4 percent to 8.4 percent, and in other cities from 6.9 percent to 8.7 percent. This finding makes sense given the privatization of Telasi, the increase in tariffs, and a shift to more expensive LPG in other major cities.

Expenditures on electricity were significantly higher in Tbilisi than in rural areas, consistent with Tbilisi's higher tariffs and far more reliable service quality. By the fourth quarter of 2001, 94 percent of households in Tbilisi received 24 hours of uninterrupted electricity, compared with 25 percent of households in other cities and only 7 percent in rural areas.[15]

Despite rising electricity prices, the absolute value of expenditures on energy fell slightly in real terms over the period.[16] One explanation is a reduction in the amount of energy used by households; another is substitution for less expensive fuels.

Changes in Service Quality

Improvements in service quality are the most direct positive effect of reform for residential consumers and the only substantial compensation for increasing tariffs. As with Armenia, the question of whether service quality has improved is therefore important, since it is an indicator of better welfare and a measure of whether reform has been successful. A reasonable proxy for service quality is the hours of service that consumers receive. In fact, most focus group participants noted that service quality had improved significantly since Telasi's privatization.

Welfare Implications of Changes in Electricity Consumption

Although prices increased and customers paid a larger share of their electricity bills, mean household consumption remained constant at about 125 KWh per month, and median consumption at about 113 KWh.[17] This reinforces the comments of focus group participants: gas use rose not because it was a substitute for electricity, but because it was a substitute for wood. Households limit their use of electricity due to cost, the obligation to pay, and to periodic supply limitations.

The findings about mean consumption have two important policy implications: first, current consumption levels were low relative to what might be expected in urban areas in a country at Georgia's level of development. Average consumption of 125 KWh per month represents extremely limited use of electricity, possibly lighting and a modest number of appliances. Electricity is certainly not used extensively for heating or air conditioning.[18] Second, in Tbilisi, where service was quite reliable, demand remained at about the same level despite price increases. This suggests inelastic demand; though prices rose, it was extremely difficult for people to respond by lowering consumption any further. This suggests large welfare losses from future electricity price increases.

An electricity demand function is typically kinked. The curve slopes steeply around the minimum required for basic needs, since few households consume less than this threshold amount. Demand in this part of the curve is inelastic. The curve then rapidly levels off as the quantity of electricity consumed moves from necessity to luxury, at which point demand is very elastic. Identifying the location of the kink is important. If prices rise above this point, consumption is pushed into the inelastic part of the demand function, where consumption is already very low and welfare losses associated with rising prices are large. At prices below this point, demand is more elastic and welfare losses are smaller. The distribution of annual household electricity consumption indicates that households were most likely to consume between 875–1,750 KWh per year (figure 5.4). A separate estimate of the demand curve confirms that households in Tbilisi were consuming close to basic minimum needs, that demand was inelastic, and that any future price increases would result in large welfare losses.

More Use of Gas

Most focus group participants expressed a desire to use gas, preferring it to other fuels for both cooking and heating, and to some extent for water heating. They noted that gas was cheaper than electricity, and cleaner and

Figure 5.4. Most Households in Tbilisi Consumed 875–1,750 KWh a Year, 2002
(distribution of electricity consumption)

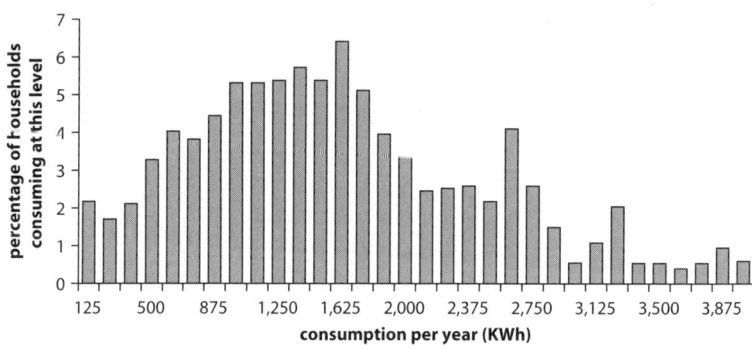

Source: AES Telasi.

more comfortable than kerosene and wood. Almost all participants without a gas connection said that they used kerosene or wood for heating and cooking, but after getting access to gas they gave up these fuels. Many said that they disliked both kerosene and wood so much that they used them only when no other option was available. Access to gas gave them a desirable substitute.

Installing a gas connection did not affect the level of electricity consumption, either because households were already managing the use of electricity to reduce bills or because the areas where they live have electricity supply restrictions. In some areas with old and nonworking meters, people were not paying for electricity they consumed, so they had no incentive to reduce consumption.[19]

Despite the obvious benefits of gas, there are barriers to obtaining it, mainly the costs of installation, the meter, and the equipment and technical difficulties in some areas.

Impact of Increased Use of Traditional Fuels

Among the key anticipated impacts of reform was that higher prices for clean network energy would increase the use of traditional fuels (wood and kerosene) by the poor. The correlation between illness and household use of traditional fuels in poorly ventilated homes is well established. The study found, as noted above, that households in Tbilisi have shifted to clean fuels, largely because of increased supply of clean and inexpensive natural gas (figure 5.5).[20] Statistical analysis of the relationship between health outcomes (such as the incidence of acute respiratory infections)

Figure 5.5. Tbilisi Households Shifted to Cleaner Fuels
(energy expenditure shares)

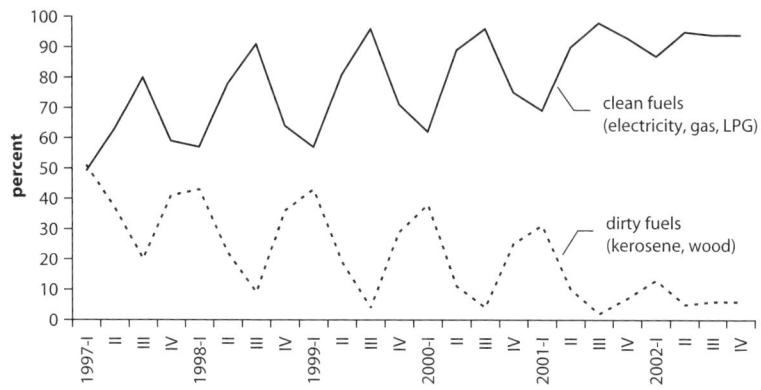

Source: Georgia Household Budget Survey.

and fuel use did not reveal the same significant correlations picked up in larger time series data sets, possibly because of the large number of confounding factors associated with observed health outcomes.[21]

In rural areas, traditional fuel consumption potentially poses a public health risk.[22] The Save the Children survey indicates that 80 percent of rural energy consumption in the winter of 2001 was fuel wood. Many other variables influence wood consumption, including forest cover, access to other fuels, proximity to forests, the availability of household labor to collect firewood, and temperature. There may be welfare gains from increasing access to cleaner more efficient wood burning technology, which could reduce the cost per effective Btu, though it remains to be determined whether households would adopt these technologies.

How Was the Utility Able to Increase Revenues?

After major efforts to increase payment collections, AES Telasi dramatically improved its revenues, increasing receipts by 135 percent by 2002.[23] While tariff increases accounted for some of the increase, better collections from customers and increases in the volume of government transfers to consumers also played a role. AES Telasi was particularly successful in reducing household arrears, with collection rates rising from 44 percent in 2000 to 86 percent in 2002.[24] AES data suggest that metering and subsidies had a much larger impact on collection rates and revenues than service quality and retail prices (table 5.1).[25]

Table 5.1. Aggregate Impact of Reform on Collection Rates in Tbilisi

	2000	2001	2002	Change 2001 over 2000 (percent)	Change 2002 over 2001 (percent)
Telasi received power (million KWh)	2,790	2,380	1,200	−15	−6
Telasi requested power (million KWh)	3,230	2,760	1,290	14	−20
Ratio of received to requested power (percent)	86	86	93	0 percentage points	7 percentage points
Average price (GEL/KWh)	0.093	0.100	0.124	8	24
Households remetered (percent)	38	69	76	32 percentage points	7 percentage points
Consumption (million KWh)	2,350	2,310	2,490	−2	24
Billings (million GEL)	217	232	309	7	33
Total receipts (million GEL)	96	186	266	93	44
Subsidies (million GEL)	35	44	55	25	26
Winter Heat Assistance Program (million GEL)	29	37	47	28	27
Government privileges (million GEL)	6	7	8	11	21
Payments by customers (million GEL)	61	142	211	132	49
Collection rate from households (percent)	44	80	86	36 percentage points	6 percentage points

Source: AES Telasi.
Note: Table includes only Tbilisi households in the sample. Requested and received power in 2002 is only from January to June.

To learn more about how Telasi raised collections from households, the study conducted a multivariate analysis to disaggregate some of the important factors driving payments. The analysis estimated receipts as a function of service quality (ratio of requested and received energy), price, enforcement (percentage of households remetered), and subsidies. It also controlled for monthly temperature and the temporary loss of thermal power plants in the winter of 2001.[26] It was complemented by focus group discussions that solicited the views of Telasi customers on a wide range of payment-related issues, including remetering, enforcement, and service reliability. These tools can help answer whether improvements in service quality make people more likely to pay their bills, how remetering increases collections, and whether nonpayment is due to affordability or free-riding.

Prices

It is difficult to untangle prices from enforcement and service quality in improving Telasi's revenues. Higher prices would be expected to increase revenues, but price increases could also reduce consumption and increase nonpayment.

The simple tabulations in table 5.1 indicate that consumption and collection rates increased as tariffs increased. Remember that consumption depends not only on price, but also on electricity supply (increasing during this period) and demand.

Subsidies

AES Telasi's revenues from subsidies grew in absolute terms, largely due to the increasing Winter Heat Assistance Program (WHAP) benefit, funded and administered by the U.S. Agency for International Development (USAID).[27] The program accounted for 29 percent of household receipts in 2000, and about 18 percent in 2001 and 2002. In addition, government privileges accounted for anywhere between 3 percent and 6 percent of AES Telasi's yearly receipts. Subsidies fell as a share of revenues because of the large increase in collections from households.

Service Quality

Since AES emphasized service quality as the positive effect of reform for consumers, it would be interesting to see whether improvements in service quality encouraged people to pay their electricity bills. It was not possible to study in detail how changes in hours of service affected individual payment rates and arrears because the data needed to relate aggregate hours of supply within the AES Telasi service area to hours of service for individual customers were not available. But it was possible to compare collection rates with the ratio of received to requested power. No substantial correlation was found, possibly because Tbilisi receives close to 24 hours of service a day.

Reliability of supply did not seem to be a major direct factor affecting the electricity payments of focus group participants. But some noted that they were anxious to get new meters because "supply is better when you have them." Some participants also expressed dissatisfaction with Telasi's failure to adhere to its original promise that if customers paid their bills, they would have 24 hours of improved electricity service. This suggests that service quality may have affected the payments of some households and supports the argument that increased tariffs and collections need to be explicitly linked to service quality and supply improvements.

Remetering and Enforcement

To improve payment enforcement, AES Telasi invested US$60 million installing electricity meters in Tbilisi. In the statistical analysis, enforcement explains much of the improvement in collections. With remetering a proxy for enforcement, collection rates are systematically higher for remetered households.[28] There is no statistically significant difference in consumption between remetered households and those having old meters, but remetered households pay a systematically higher percentage of their bill at all consumption levels—on average twice as much as those not yet remetered—and their arrears are significantly lower.

The multivariate analysis did not tell whether metering facilitates cutoffs for nonpayment (enforcement) or adds credibility to the invoice (a proxy for service quality). So the interaction between metering and payments was a key issue addressed in the focus groups. The responses of participants indicated that metering plays a complex role. Participants feared supply cutoffs, controlled consumption, and trusted that the amount of their bills was accurate (though some expressed doubts about the accuracy of the new meters, which appeared to "go faster"). Some also noted the advantage of reduced corruption due to the new meters, though others saw this as a disadvantage.

The fear of cutoff was particularly strong—even though Telasi said that it probably cut off only 10 percent of nonpaying households in each month. This suggests that the threat of disconnection (particularly if likely at an inconvenient time) may be almost as effective as an actual cutoff in reducing nonpayment. The risk of disconnection was also cited as a factor in installing an illegal connection. Participants who paid their bills expressed resentment that Telasi did not do a better job tracking down and removing illegal connections.

Nonpayment: Affordability or Free-Riding?

Improving collections could have a disproportionate impact on low-income households. But changes in collection rates by income class indicate that they increased uniformly across the lowest and highest 20 percent of households, suggesting that free-riding rather than affordability was behind the arrears. If affordability were more important, collections would be lower for the lowest 20 percent.

After experimenting with alternative approaches to metering, AES and the management contractor for distribution outside Tbilisi concluded that communal metering—where a community bears collective responsibility for paying the bill—can be more effective than individual household meters for increasing collections and keeping costs down. In part this was because the threat of cutting off a whole neighborhood encouraged better self-policing within communities. With the cost of metering at around US$75 per meter installed, this was also a far more cost-effective means of improving collection rates.

How Effective Were the Mitigating Transfers?

Recognizing the need to mitigate the impact of rising prices on the poor, Georgia has a range of programs providing energy transfers to households. One provides all veterans and pensioners with a set allocation of electricity every month.[29] Refugees and internally displaced persons also receive substantial quantities of free electricity, while other programs provide certain households with 850 cubic meters of natural gas per year.[30] In addition to the government-funded transfers, the WHAP has been providing families with US$12–$35 worth of electricity a month.

One of the most contentious debates among those working on power sector reform is tariff-based subsidies (such as lifeline tariffs) versus direct income transfers. One of the most useful characteristics of the AES utility data is that the bills identify whether the households received government transfers or the WHAP. Merging these data allows the identification of energy consumption and payment patterns by welfare group and measures the targeting effectiveness of the transfers.

The percentage of households receiving the government transfer—paid to veterans and pensioners, and not specifically poverty targeted—was evenly divided across all quintiles (table 5.2). The WHAP transfer, which is poverty targeted, accrued more to households in the lower quintiles, but a significant share of the WHAP went to households in the high expenditure quintiles in 2000 and 2001.

Table 5.2. Electricity Subsidy Incidence by Quintile in Tbilisi

	Percentage of households receiving government subsidy				
	Income quintile				
Year	Lowest	Mid-Low	Middle	Mid-High	Highest
2000	12	12	15	13	13
2001	10	16	18	11	10
	Percentage of households receiving Winter Heat Assistance Program subsidy				
2000	25	16	18	17	10
2001	27	23	21	19	14

Sources: Georgia Household Budget Survey and AES Telasi.

How much consumption was covered by the transfers? Recipients of the government transfer received 27–32 percent of their annual electricity for free (table 5.3), and WHAP transfer recipients received 54–64 percent of their electricity free. More detailed analysis suggests that WHAP recipients did not necessarily use the free electricity for heating—in many cases, they used the subsidy to smooth their consumption through the entire year. This may explain in part how households maintained (and sometimes even increased) electricity consumption despite substantial tariff increases.

A large share of government transfers for electricity are compensation for electricity consumption beyond levels that might be considered "basic"—suggesting that the government, in many instances, financed consumption in excess of what typical households would be willing to consume if they were obliged to pay from their own household budgets. The welfare gains of providing large electricity transfers to households (amounts greater than 150 KWh a month) were probably small.

Transfers were also a major contributor to government expenditures on the electricity sector, which between 2001 and 2003 increased from 43 million GEL to 98 million GEL—or 7.3 percent of total government expenditures. Some of this increase was due to monetizing formerly hidden subsidies and to higher government expenditure on electricity in the public sector, but the higher cost of electricity transfers also contributed (table 5.4). Meanwhile the cost of subsidies for gas supply, provided by both the state and municipal budgets, was also rising as additional households eligible for subsidized gas connected to the gas network.[31]

Because transfers are generally intended as a tool for alleviating poverty and increasing equity, the merits of the current system are ambiguous. While the welfare gain to households associated with the misdirection of transfers is small, the burden on government expenditures is large.

Table 5.3. Transfer Coverage in Tbilisi

	Government subsidy		USAID subsidy (Winter Heat Assistance Program)	
	Mean annual (KWh)	Percentage of KWh free	Mean annual (KWh)	Percentage of KWh free
2000	1,851	28	1,440	54
2001	1,659	32	1,461	64
2002	1,948	27	1,691	56

Sources: Georgia Household Budget Survey and AES Telasi.

Table 5.4. State Budget Payments to the Energy Sector, 2001–03
(thousand GEL)

Name	2001	2002	2003 (Planned)
Direct subsidy to the Ministry of Fuel and Energy	3,000	13,000	36,500
Reimbursement of the fee for electricity consumed by the refugees	6,555	13,646	14,016
Reimbursement of the fee for electricity consumed by the budget organizations (central, local)	21,924	27,346	29,348
Sums allocated for energy sector through special decrees	6,000	10,160	4,500
Compensation for the various categories of population	2,800	3,000	11,500
Total direct support	42,280	69,154	95,864
Total budget expenditures	**906,314**	**1,031,259**	**1,343,700**
Energy sector support/total budget (percent)	4.7	6.7	7.3
Foreign credits and cofinancing	**17,279**	**34,325**	**46,500**

Source: Ministry of Finance.

The government's poor fiscal situation and the competing demands for social transfers elsewhere represent a major motivation for remedying this situation.

Proposing a Better Mitigating Strategy

Using the data from the HBS and AES Telasi, the authors simulated the targeting and cost-effectiveness of an alternative transfer design given to households whose consumption falls within a certain margin of usage. Ideally, the targeting would be based on a rolling average of household consumption—say, in the three previous months. But because there is surprisingly little differentiation in consumption between the lowest and highest 20 percent of households during the summer months, a simple simulation performs better if based on annual consumption (figure 5.6). The proposed transfer would be given to households consuming between 875–1,750 KWh per year—or between 73–145 KWh per month. The lower bound is set to exclude residences not regularly occupied, summer houses, for example. It also eliminates incentives for gaming the system, for example, by installing multiple meters.

The simulation showed that the proposed transfer would reach a higher percentage of low-income households than either of the existing

Figure 5.6. Household Electricity Consumption in Tbilisi

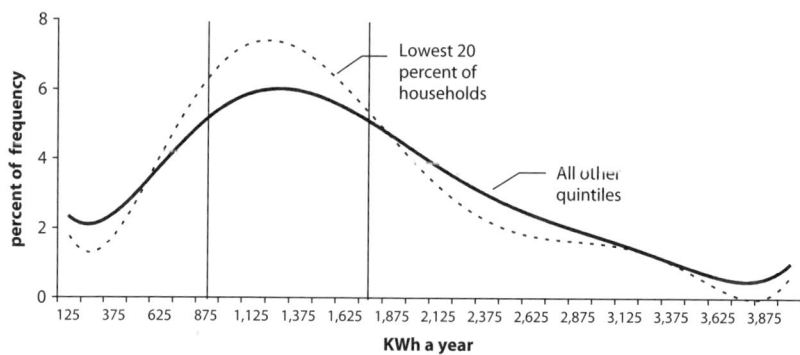

Sources: Georgia Household Budget Survey and AES Telasi. (See box 5.1).

transfers (table 5.5). It would also reach a higher percentage of the other quintiles. The absolute transfer to each household would be substantially lower than in either of the existing programs, and the total cost would fall between the WHAP and the government program. More cost-effective, the alternative would thus save the government money overall.

Three important caveats apply:

- First, the total cost of transfers would increase as the old transfer is phased out, since more households are likely to consume in the 75–125 KWh range (though poverty targeting may well improve as the old transfer is phased out, and with the loss of the existing transfers, electricity consumption will be based more directly on actual household income).

- Second, several well-organized stakeholders could encourage the government to keep the current system in place, including veterans, who do not wish to lose their benefits, and the utilities, which presumably enjoy the simplicity and predictability of payments associated with the current system.

- Third, these results are based on data from Tbilisi, so caution is needed in generalizing them to the rest of the population.

One possibility would be to pilot the alternative assistance program. The HBS could be linked directly to the utilities' billing and payment databases to monitor the poverty targeting of the transfer. Over time, as data on consumption patterns, income, and payments are collected and analyzed, the targeting system could be further refined to reduce overall costs.

Table 5.5. Simulation of Cost-Effectiveness of Different Transfers in Tbilisi

	Quintile				
	Lowest	Mid-Low	Middle	Mid-High	Highest
Households receiving (percent):					
Government subsidy	10	16	18	11	10
Winter Heat Assistance Program subsidy	27	23	21	19	14
Proposed subsidy[a]	44	38	40	42	39
Proposed subsidy—no gas users[b]	40	35	43	34	35
Average subsidy per household (KWh a year)					
Government subsidy	610	561	548	646	535
Winter Heat Assistance Program subsidy	1,000	1,000	1,000	1,000	1,000
Proposed subsidy[a]	407	411	497	476	324
Proposed subsidy—no gas users[b]	398	384	479	382	287
Cost-effectiveness (GEL per household)					
Government subsidy	76	70	68	80	66
Winter Heat Assistance Program subsidy	124	124	124	124	124
Proposed subsidy[a]	50	51	62	59	40
Proposed subsidy—no gas users[b]	49	48	59	47	36

Source: Lampietti and others 2003.
a. The proposed subsidy is for households that consume between 875 KWh and 1,750 KWh a year. These households are given a monthly subsidy equal to the difference between 125 KWh and their monthly consumption. The assumed tariff is 0.124 GEL/KWh.
b. The second proposed subsidy is identical to that described in note a, except that it is available only for households that do not have access to natural gas.

Conclusion

This study was conceived to understand the dynamics of nonpayment and to move forward with the reform process in Tbilisi and elsewhere. By the time the study came out, electricity sector reform was in further disarray, with AES on the verge of withdrawing from Georgia. In late 2003, AES sold Telasi to Russian utility RAO UES, and the Rose Revolution transformed the context for reform.

Even so, this study highlights the problems of the reform program and nonpayment. The multivariate results indicate that remetering and price are equally important determinants of receipts, followed by service quality and social benefits. Remetering, in conjunction with tariff increases, should be a high priority, particularly in the early stages of reform, to generate the maximum amount of revenue. If investment capital is limited, communal metering can be even more effective than individual metering—though the latter is valuable in implementing an effective mitigating strategy.

The data also suggest, in Georgia at least, that an aggressive approach to reducing nonpayment did not necessarily have a disproportionate adverse impact on low-income households—particularly if suitable subsidy and transfer mechanisms could address cases of severe need. The study simulated an alternative subsidy and provided empirical justification for adopting it.

This study also illustrates the importance of institutional and political economy factors in improving cost recovery. With an endemic propensity of electricity consumers to not pay, it was unrealistic to hope for a simultaneous increase in collections and tariffs in Georgia. But the institutional backdrop left AES at a fundamental disadvantage. Such an ambitious reform agenda as Georgia's cannot work without a strong regulator and a willingness within the sector to play by those rules—all shown to be lacking in Georgia. As a later report put it, "Factors that inhibited a better outcome include political pressure in the operation of energy companies to provide electricity at any costs; strong vested interests to maintain the status quo; theft; corruption; political tolerance of nonpayment; lack of incentives on the part of corporate management to resist political pressure and vested interests; and weak enforcement of laws and regulations."[31] Whereas a dramatic improvement in Armenia's payment levels was commonly ascribed to increased government commitment to improving cost recovery, Georgia during the AES years showed how a lack of high-level political commitment can hurt the reform process.

The government's failure to back the rules of the game was exacerbated by flaws in the design of privatization. Though paying customers were promised full electricity supply, AES Telasi was dependent on an intermediary, the Georgian Wholesale Electricity Market (GWEM), to ensure those customers who paid were supplied. But with GWEM subject to political interference, this arrangement prevented AES Telasi from creating an effective incentive regime for payment and undermined the important link between increased payments and service quality improvements. In addition, the government collected value-added tax (VAT) payments from AES based on the quantity of electricity distributed rather than the quantity actually paid for, meaning it had little financial incentive to back AES efforts at improving payments. If investors are to be found and maintained in the future, not only must efforts be made to increase collections before privatization (as in Armenia), but privatization contracts must ensure that all players are united by the same incentives. The interests of the government and the utility clearly need to be aligned to back reform and share the risk of nonpayment.

Unsurprisingly, AES's experience in Georgia has become a celebrated case study for energy sector privatization. The lessons, also examined in business schools, informed the management contract experience of the United Energy Distribution Company (UEDC), the utility responsible for distribution outside Tbilisi. UEDC managers concluded that management and proper control are essential for the success of all other activities. Under a management contract, the government is on the same side of the table as the utility, government support becomes more reliable, and the foundations can be laid for viable privatization.

Although the change in government renewed the commitment to improving financial viability in the sector and improving supply, a 2005 World Bank report cited continuing nonpayment, accumulated debts, theft, and possibly corruption as reasons why Georgia's energy sector remains financially bankrupt. The GWEM is now managed by a private consortium, and under a management contract UEDC has improved supply and dramatically increased collections outside Tbilisi (without a proportionate increase in social transfers). But low collections still thwart much-needed investments, and the sector's performance continues to hold back economic growth.

Georgia spotlights the difficulties encountered by utilities in pushing for cost recovery in a hostile environment. The pace of reform was fairly fast, and the government rapidly initiated all reform measures and overtly supported AES Telasi. But then the government undermined the utility's efforts at increasing revenues, and exploited resentment toward a foreign operator to insulate itself from the political fallout of reform.

This study, by showing that nonpayment in Tbilisi was due to freeriding rather than affordability, suggests that despite the protests, collections could be increased without necessarily hurting the poor. The next case study shows that the studies can go a step further, exploring the reality behind a fierce debate on reform and privatization and the poor—in Moldova.

Notes

This chapter is based on Lampietti, Gonzales, Hamilton, Wilson, and Vashakmadze 2003.

1. World Bank (1999b, 2002e). The 2002 report concluded that since 1996, poverty had increased steadily, average consumption had fallen, inequality had risen, and living standards had declined. In real terms, average monthly per capita expenditure fell 4 percent from late 1996 to late 2002. See World Bank (2005a).

2. World Bank (1997a).

3. The same pattern occurred in the gas sector, though most gas was imported from Russia. Gas distribution companies in urban centers outside Tbilisi were sold to Sakgazia, a joint venture between local partners and the Russian gas supply company, Itera. Tbilgazi, the gas distribution company serving Tbilisi, was offered for privatization on a number of occasions, but the only credible bidder was Itera. The government regards Itera's ownership of Tbilgazi as an undesirable step toward the vertical reintegration of the gas supply sector, so Tbilgazi remains a municipally owned utility.

4. Some of the smaller electricity distribution companies were sold to local investors. Eight small companies (less than 5 percent of the market in total) in the Kakheti region have been sold. Two hydroelectric plants (Khrami I and Khrami II) were also given to AES under a 25-year concession, and after protracted negotiations the bulk of Georgia's thermal generation capacity was sold to AES in April 2000.

5. World Bank (1999d).

6. This situation originated in the Georgian Wholesale Electricity Market (GWEM), established in 1999 to manage transfers between different electricity enterprises and transfer electricity payments from distribution companies to generation companies. Rather than directly distributing the electricity that it was generating, AES sold the electricity it generated to GWEM. AES Telasi would then buy electricity for distribution from GWEM. The problem occurred because GWEM would not make the electricity available to AES Telasi. In the end AES Telasi bypassed GWEM to ensure that it received the electricity it was paying for.

7. External arrears reduced the ability to import electricity from such neighboring countries as Armenia. Prolonged drought reduced the availability of hydroelectricity, and an explosion at the Gardabani thermal plant reduced thermal generation by half for much of the winter of 2001.

8. Save The Children (2002).

9. Tbilgazi's customer base increased from 39,000 households in June 2000 to 164,000 households in January 2003. There are approximately 300,000 households in Tbilisi. In the HBS, households were asked if they had a natural gas connection. These data indicate that the number of connections decreased nationwide from 1998 to 2000, with a small increase in Tbilisi from 2000 to 2001.

10. Gas supply was intermittent, though it appeared to be stabilizing as external arrears were paid off. Gas was purchased from the Russian company Itera by industrial customers, from the Gardabani power plant, and from the local gas distribution companies. In the past, Itera tied gas delivery to payments from

any and all of these large customers. So if one or more of them accumulated significant arrears, gas supply to the country was curtailed until a satisfactory settlement could be reached. The completion of the Baku-Tbilisi-Ceyhan pipeline was expected to further reduce supply constraints by providing an alternative to Russian gas imports, though events of January 2006 suggest that Russian control of gas supply remains a very sensitive issue (BBC News Online, January 22, 2006).

11. Prices are in nominal terms to reflect tariff increases, including those imposed by the reform.

12. The prices are weighted national averages, which are based on data taken from the quarterly HBS. These prices are in cost per unit of effective energy output, rather than the prices that customers pay per unit of energy input. The adjustment was based on typical conversion efficiencies of the fuels and the efficiency of different types of appliances. This implicitly assumes that all households have the same technology.

13. Gas tariffs at the end-user level cover the cost of importing the gas from Russia (approximately US$60 per 1,000 cubic meters), transmission charges, and the costs of local distribution. The transmission and distribution margins have been reviewed regularly by GNERC (Georgian National Energy Regulatory Commission), and the companies are entitled to apply for a tariff increase based on demonstrated costs of service supply.

14. This is not necessarily because of the reform program. For example, the large jump in the kerosene price in 1999 may be related to a rise in international crude oil prices, which rose from US$10 a barrel to US$22 between January and September.

15. Households in the HBS were asked to report the number of hours of electricity received during the week prior to the interview. Households were asked this question only during the first of four interviews. The results shown here are for the quarter in which the initial interview took place.

16. Household fuel expenditures are converted into physical units (million Btu) by dividing expenditures by unit price per million Btu and adjusting the physical units to reflect the conversion efficiencies of typical energy-consuming appliances.

17. The data set contains a large number of zeros during the first few months of 2000, so the median is close to zero. One explanation is that the billing system started in the middle of 1999, so the large number of zeros is part of the adjustment period during the creation of the data set. A second explanation is that there were few existing meters in the system during this period. Before new meters were widespread, an "average" or "estimated" amount of KWh was assigned to households as their consumption. These numbers were later

verified by AES Telasi as new meters were introduced into the distribution system, sometimes resulting in very large bills for the households.

18. A refrigerator (manual defrost 5–15 years old) consumes about 95 KWh per month, and three incandescent lightbulbs consume another 30 KWh per month.

19. According to information from AES Telasi, in some areas estimates show that supply accounts for GEL 60–70 per household a month while payments are only GEL 2–3 per household a month.

20. This pattern holds for the lowest 20 percent and for the average household.

21. According to the Save the Children survey, in 2002 more than 53 percent of households had one or more members with a chronic disease, and 76 percent of households had one or more members with either an illness or disease in the previous three months. It is therefore possible that other factors for which there are no data mask health differences related to fuel use. For more details on time series data, see Lampietti and others (2003).

22. The health impact will depend on the number of households and the technologies used when burning traditional fuels—for example, improper stoves.

23. Revenue from the residential sector increased 91 percent from 2000 to 2001 and another 41 percent from 2001 to 2002. These figures are for a sample of 1,349 households included in the Georgia HBS. In total, AES Telasi has approximately 300,000 customers. Households participating in the HBS were randomly selected and may be presumed representative of households in Tbilisi.

24. At times collection rates even exceeded 100 percent of current billings, as households settled arrears and transfer payments for subsidies were received from U.S. Agency for International Development (USAID) or the government. Arrears for public sector customers were another very important issue.

25. The cost of meters is not taken into account in this analysis.

26. The program finances the supply of electricity to low-income households in Tbilisi for winter heating during the January–April period. The amount each household receives has varied each year depending on the funding available. It was 850 KWh in 2000 and 1,000 KWh in both 2001 and 2002. The planned amount for 2003 was 480 KWh.

27. Remetering refers to both replacing old meters for newer ones and installing meters outside the dwelling; households used to have meters inside the dwelling.

28. Before 2003, this was between 35 KWh and 70 KWh per household a month; it later increased to 240 KWh a month in the winter and 120 KWh a month in the summer.

29. This program is part of the "President's fund," which covers veterans.

30. Gas connections in Tbilisi increased from 10,000 households in 1996 to 170,000 in 2003.

31. World Bank 2005a, p. 60. The report continues to state that this was compounded by "[a]n increasing fragmentation of political power during the Shevardnadze government [that] reduced high-level commitment to curbing corruption and advancing difficult reforms" (p. 77).

CHAPTER 6

Does Privatization Hurt the Poor of Moldova?

By the end of the 1990s, the perception was widespread that reform, particularly privatization, hurt the poor. While AES Corporation was on the brink of withdrawing from its Georgia operations, the international financial institutions and another foreign private utility operator in Moldova were confronted with a government threatening to reverse privatization, using the popular perception of privatization's negative effects on the poor to justify its actions. This study shows how understanding the effects of reform on households can bring clarity and empirical evidence to debates that are ideologically motivated and highly politicized.

The Long Slide

Moldova's postindependence decline lasted the whole of the 1990s, leaving it one of the poorest countries in the region.[1] The war in Transnistria in the early 1990s weakened central government control over an area that holds much of Moldova's industrial and power generation capacity.[2] And transition, poor governance, and corruption took a heavy toll on the economy. With declining economic opportunities and rising poverty, up to a quarter of Moldova's 4.2 million population is estimated to have left the country in search of work.[3]

Moldova depends on outside energy sources, importing more than 95 percent of its energy from the Russian Federation and Ukraine.[4] The movement of previously low Russian and Ukrainian gas and oil prices toward international levels contributed to the rapid accumulation of debts by the state energy company, Moldenergo, US$300 million by 1995.[5] Until 1998, residential energy tariffs remained low, and sector revenue could not cover the cost of imports. Funds for maintenance and repairs dried up, decapitalizing power infrastructure assets. By the late 1990s, Moldova was facing an energy crisis. Cash-flow problems made Moldenergo vulnerable to supply shortages, resulting in regular power interruptions and lower quality. The areas outside Chisinau, where many poor people live, were hardest hit by rationing, with many localities receiving electricity for just a few hours a day.[6] Power was often interrupted without warning, and per capita monthly electricity consumption plunged to the lowest levels in Europe, (figure 6.1), at just 51 kilowatt hours (KWh) in 2001—a quarter of the average in the Europe and Central Asia (ECA) region and less than half the basic minimum need.[7]

In 1997, Moldova launched a reform program, and in 1999 tariffs were increased 84 percent, followed by smaller increases.[8] In 2000, the government adopted a law on nominative targeted compensation (NTC) for energy use to help vulnerable groups cover the rising cost of their energy consumption.

Also in 2000, part of the distribution network was privatized. Three of five regional electricity distribution companies (REDs)—RED Chisinau (serving the capital region), RED Centru (serving central Moldova), and

Figure 6.1. Electricity Consumption in Moldova Plunged between 1992 and 2000
(total annual electricity consumption)

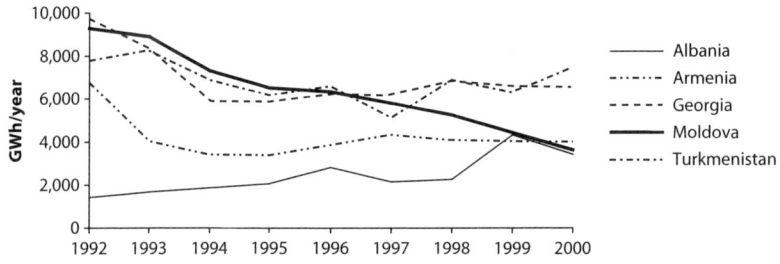

Source: International Energy Agency.

RED Sud (serving southern Moldova)—were sold for US$26 million in an open tender to the Spanish utility Union Fenosa, which as part of the deal committed to invest US$56 million in infrastructure rehabilitation over five years. The Union Fenosa service area covered 694,000 residential and 33,000 nonresidential customers (60 percent of Moldova's population).[9] Two other regional distribution companies, together known as the NREDs, remained state owned.

After the peak of the energy crisis in 2000, reform produced substantial improvements in supply to consumers. But reform and privatization elicited acrimonious debate among stakeholders and questions about the costs and benefits of reform. Much disagreement centered on Union Fenosa, which became the country's largest foreign investor in a highly visible privatization deal. Union Fenosa electricity tariffs were about 10 percent higher than those of state companies, fueling concerns that the profit motive left consumers, particularly the poor, worse off. Union Fenosa covered only 38 percent of its investment commitments from 2000 to 2002, a failure that it ascribed to uncertainty in the investment climate, including a lawsuit centered on irregularities in the privatization procedure and the government's reluctance to allow further tariff increases.[10]

Moldova's government, which in 2001 became the first explicitly Communist government to be elected in a post-Soviet state, added to the uncertainty by openly announcing its intention to reverse privatization, including privatization in the energy sector. Other countries saw similar debates over the pros and cons of privatization, but in Moldova, the government's unambiguous agenda of reversing privatization was particularly pressing.

This study sheds light on a very contentious debate by providing empirical answers to two intentionally neutral questions. First, did reform affect the poor and the nonpoor differently, as was charged by opponents of reform? Second, were household electricity consumption patterns different in private and public distribution networks?

Box 6.1

Data for the Analysis—Moldova

To determine whether the impact of reform was the same for poor and nonpoor, the study compares three quantitative welfare indicators: electricity consumption, electricity expenditures (payments), and share of electricity expenditures in total

(Continued)

household expenditure. To determine how privatization affected consumers, it compares customers served by Union Fenosa and the NREDs. It therefore provides a counterfactual view of privatization by comparing consumer outcomes in private and public regions.

The quantitative analysis relies primarily on time series data from the Moldova household budget survey (HBS) and records provided by Union Fenosa. The HBS is a survey of more than 6,000 households conducted annually since 1997. Data from the survey were compared with data provided by Union Fenosa to test the reliability of survey responses to questions about electricity consumption, billing, and payment. For NRED customers, only aggregate, not household utility data, were available. But the two sets of data from the HBS and Union Fenosa were highly correlated, increasing confidence in the HBS data for NRED customers.[a] The HBS data were used to estimate a household electricity demand function and compare price elasticity of demand by for different income groups.[b]

Since the quantitative data were already available, the qualitative analysis was conducted afterward, to confirm and improve understanding of key questions emerging from the household data analysis. The qualitative analysis is based on focus group and key informant interviews in the winter of 2003–04. Forty-three focus groups and 59 key informant interviews were held with poor and nonpoor people, living in large cities, small towns, and rural areas, with access to different sources of energy, living in areas served by Union Fenosa and the NREDs. Interviews were also held with distribution company managers, meter readers, postal workers, social assistance providers, and mayors.

The analysis covers the four years starting with 2000 and ending in 2003. Limiting the analysis to this period carries four important caveats. First, although Union Fenosa took control of part of the network in February 2000, reform began in 1997, and the largest tariff increases occurred before 2000. So the quantitative analysis does not capture the largest price effects on household welfare. Second, the psychological point of reference for most people is the early 1990s, when Moldova was more prosperous and reliable electricity was virtually free. This tends to bias their responses to questions about the recent privatization, because they do not compare it to the mid-1990s when the system was close to collapsing. Third, the recent growth in the economy coincides with privatization, which complicates inferences about the impact of reform on households. Fourth, the high level of emigration in search of employment that began during the 1990s also introduces uncertainty about aggregate consumption figures and may have had a disproportionately large effect on the number and size of poor households.

(Continued)

Another caveat is the dramatic decline in poverty during this period and how this may affect the analysis. Between early 2000 and late 2003, Moldova's poverty level fell from 71 percent to 37 percent, with the greatest decline in rural areas. Studying the same households over the four years (using "panel data") would have enabled the study to analyze the consumption changes of households that started poor and joined the nonpoor. Instead, the study was only able to look at aggregate data for poor and nonpoor groups. This limitation in the data means that the findings understate changes in consumption for the poor. Households originally "poor" increased their consumption as they became "nonpoor." But this increase in consumption is not captured in the poor group where they started, since they are part of the nonpoor category the next time their consumption is measured.

Source: a. Relying on the HBS also allowed the study to use the same definition of poverty as did the World Bank poverty assessment (World Bank 2004c). The poverty line is 196.03 MDL per month.
b. See World Bank (2004f) for details.

Residential Energy Consumption in Moldova

Between 1998 and 2003 the cost of all energy products increased, but electricity tariffs rose most rapidly (figure 6.2).[11] Although district heating tariffs increased even more than electricity, it is unclear whether payment for this service was enforced, given the difficulties of enforcing district heating payments and the collapsed district heating systems in most towns.[12]

Residential electricity consumption in Moldova was very low. Monthly household electricity consumption averaged 61–84 KWh between 1997 and 2003,[13] less than one-tenth of the 852 KWh average in the United

Figure 6.2. Electricity Was the Most Expensive Source of Energy in Moldova

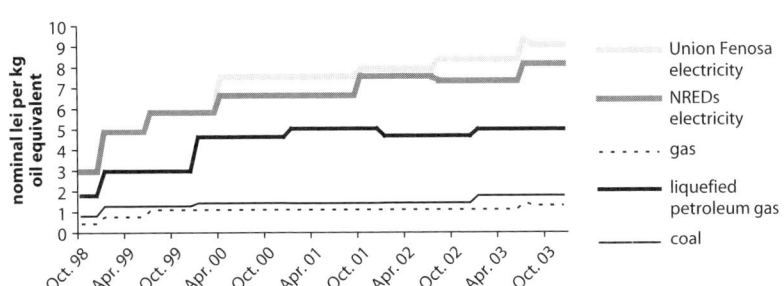

Sources: Moldova Household Budget Survey and ANRE.

States.[14] Even compared with other relatively poor countries in the region, this was extremely low. Sixty KWh a month was enough to run only a refrigerator for 5.5 hours a day and three 75-watt lightbulbs for 4 hours a day. Many Moldovas, especially the poor, were thus extremely restricted in their electricity consumption and had to cope by consumption reducing measures, such as unplugging appliances. Public and private institutions, including schools, hospitals, and cultural centers, were also unable to pay for electricity. Despite significant improvements in supply, public areas remained dark. Communal areas in apartment blocks, such as stairwells, often remained unlit when money or trust for making collective payments was lacking. And there were reports of residents suffering injuries from navigating stairwells in the dark when elevators were not functioning. Safety at night was a concern, with many urban and rural residents afraid to leave their homes because the streets were dark.

In addition to normal uses, electricity in urban areas was sometimes used as a supplement or substitute for poorly or nonfunctioning district heating. District heating served 98 percent of households in large cities and 29 percent of households in small towns, according to the 2003 HBS. Where available, households reported spending a larger share of income on district heating than on gas or electricity. It is not clear, however, whether households were reporting the amount they were billed or the amount they actually paid—nonpayment for district heating was reported to be quite high.

In small towns without district heating, heavily subsidized gas was the heating fuel of choice if it was available, followed by electricity. This pattern of use implied that future network energy price increases were likely to hit people living in small towns hardest. The focus groups revealed that rural households rarely cooked or heated with electricity, usually using wood, coal, or gas.[15] Access to piped gas is becoming more common with a government-financed program to provide gas to every area of the country by 2010.[16]

Effect of Reform on Electricity Consumption

On average, the poor consumed 26 percent less electricity than the nonpoor (figure 6.3). But since 2000, the poor increased monthly electricity consumption by 14.6 percent (from 48–55 KWh), while the nonpoor increased consumption by only 3.2 percent (from 62–64 KWh). So despite rising tariffs, the poor were catching up with the nonpoor. These findings are consistent with the household demand model, which

Figure 6.3. The Gap Narrowed in Electricity Consumption between the Poor and Nonpoor

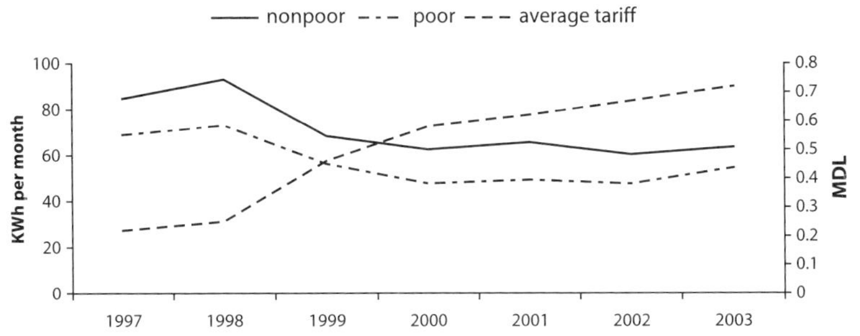

Source: Moldova Household Budget Survey, Department of Statistics, Moldova.

shows no difference in the price elasticity of demand between poor and nonpoor households.

Payment rates for the poor and nonpoor were similar, which suggests that the narrowing of the consumption gap between poor and nonpoor cannot be attributed to nonpayment by the poor. If consumption increased more among the poor than the nonpoor while payment rates reached the same levels (almost 100 percent by 2003), it follows that electricity expenditures would increase more rapidly for the poor than the nonpoor. This is exactly what is observed: in 2000, the poor spent 38 percent less than the nonpoor on electricity, but by 2003, they were spending only 18 percent less,[17] even while payment rates were the same for both groups.[18] These figures suggest that the poor were catching up with the nonpoor in electricity consumption.

In spite of increased consumption and higher collection rates, the share of expenditure on electricity declined for both poor and nonpoor. The poor continued to spend a larger share of their income on electricity than the nonpoor (4.7 percent versus 3.4 percent in 2003), but the gap was closing, as the share of income spent on electricity declined more rapidly for the poor (table 6.1).

Effect of Reform on Service Quality

Availability of electricity improved greatly, and blackouts were dramatically reduced nationwide. The poor, disproportionately affected by blackouts, benefited most from the return to 24-hour service. Findings from focus groups confirmed that the majority of Moldovans were satisfied with improved service quality and reliable supply.[19] They also reported

Table 6.1. Share of Electricity Expenditures by the Poor and Nonpoor, 1999 and 2003
(MDL a month)

	Poor			Nonpoor		
Item	1999	2003	Percentage change	1999	2003	Percentage change
Electricity expenditures	28	40	42.9	33	47	42.4
Income	431	966	124.1	871	1,799	106.5
Share of electricity in income (percent)	7.4	4.7	na	4.3	3.4	na

Source: Moldova Household Budget Survey.
na = not applicable
Note: Gross household expenditures are used as a proxy for income. Percentage changes were computed using household data, not the aggregate data in the table.

no difference in service interruptions among poor and nonpoor households after reform.

Focus group discussions also indicated that voltage levels and frequency fluctuations improved, though problems remained in some localities. Poor households had a harder time repairing or replacing appliances damaged by voltage fluctuations, and therefore derived greater benefit from a reduction in fluctuations.[20]

Differences between Urban and Rural Households

Households in large cities (lowest poverty rate) and small towns (highest) spent a higher share of their income on electricity than did households in rural areas.[21] In 2003, the average household in large cities consumed 90 KWh per month, while the average household in small towns consumed 65 KWh, and the average in rural areas just 51 KWh. Between 2000 and 2003, consumption increased by 26 percent in rural areas, compared with 11 percent in large cities and 3 percent in small towns (table 6.2). These data are consistent with the findings of the poverty assessment, indicating that poverty fell most in rural areas.

Table 6.2. Change in Electricity Consumption and Expenditures, by Location

	Percentage change between 2000 and 2003		
Item	Large cities	Small towns	Rural areas
Electricity consumption	11	3	26
Electricity expenditures	38	26	48
Share of income	−27	−32	−33

Source: Moldova Household Budget Survey. See Box 6.1

Figure 6.4. The Share of Electricity Expenditures in Total Expenditures Declined After 1999

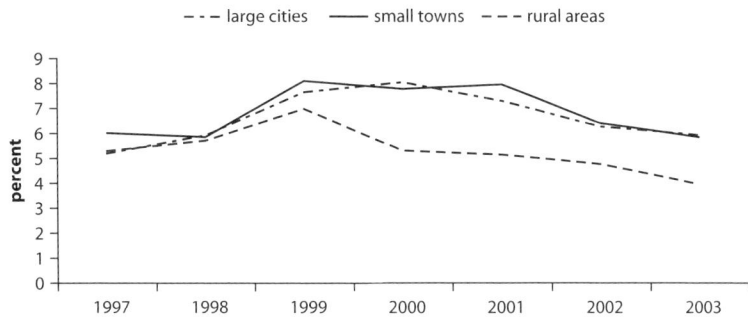

Source: Moldova Household Budget Survey. See box 6.1

The share of income spent on electricity in rural areas dropped significantly after the 84 percent tariff increase in 1999 (figure 6.4). That a drop of similar magnitude is not observed in cities or small towns suggests that rural households either went without or found less expensive substitutes for electricity.[22]

Did Reform Hurt the Poor?

Contrary to perceptions, the quantitative evidence suggests that the poor were not hurt by reform. The gap in electricity consumption between poor and nonpoor was closing, attributable not to the design of reform but to the improved electricity supply, particularly to rural areas, coupled with substantial income growth.

The qualitative analysis did reveal why people perceived that reform hurt the poor. Despite improved electricity supply and quantitative evidence suggesting that income growth offset the impact of tariffs, focus group respondents expressed anxiety over future tariff increases, consistent with a recent opinion poll that found that Moldovans were becoming more pessimistic. One explanation is that people were comparing the then-current situation with that of the early 1990s, when electricity was inexpensive and plentiful, rather than with the later 1990s. Despite very positive macroeconomic indicators, fewer respondents (26 percent) said they had a better life now than a year ago (29 percent).[23] Although the macroeconomic situation has improved significantly for the poor, Moldova's very poor remain under stress, and income growth may not have reached them. In focus group discussions, the very poor indicated that they were still a long way from raising electricity consumption to minimum basic needs. They unplug their refrigerators for days at a time, minimize the use of their television sets, and restrict themselves to low-wattage lightbulbs.

A Difference between the Private and Public Utilities?

Examination of tariff increases, disconnections, consumption patterns, and power losses reveals very little difference between Union Fenosa and the public electricity companies, refuting claims that privatization hurts the poor. Union Fenosa's residential tariffs did increase more than NRED tariffs after privatization; nominal tariffs for Union Fenosa rose 26 percent, from MDL 0.50 in 1999 to MDL 0.78 in June 2004, while tariffs charged by the NREDs rose 13 percent, from MDL 0.50 to MDL 0.70.[24] The difference is explained largely by the fact that until January 2004, the methodology for setting Union Fenosa's tariffs included a fixed return on investment.[25] Tariff methodologies then became the same for both companies and in the future will be strongly determined by the level of investment in infrastructure.[26]

Increased enforcement of electricity payments did lead to loss of access, and this was more frequent with the private sector operator. Union Fenosa reportedly disconnected 3.4 percent of its customers in 2003, the NREDs only 0.4 percent. Qualitative evidence suggests that people were often disconnected because they could not pay their bills. Reconnection fees and associated fines were often high, and consumers felt that insufficient warning time was given before they were disconnected for nonpayment.[27]

Once other factors are taken into account—tariff rates, income, household size, and apartment size—consumption patterns of households served by Union Fenosa and households served by the NREDs are roughly similar.[28] In areas Union Fenosa served, average monthly household consumption increased from 55 KWh in 2000 to 62 KWh in 2003, a 12.7 percent increase. In areas the NREDs served, consumption rose from 52 KWh to 60 KWh, a 15.4 percent increase (table 6.3). The increase in payments was also very similar. That changes in consumption were similar despite an 11 percent difference in tariffs suggests that either demand was relatively price inelastic or the NREDs had higher collections, offsetting the effect of a lower tariff on consumption.[29]

Table 6.3. Consumption, Payments, and Percentage of Income Spent on Electricity by Union Fenosa and NRED Customers, 2000–03

Item	2000	2001	2002	2003	Percentage Change between 2000 and 2003
Union Fenosa					
Average tariff (MDL)	0.62	0.66	0.7	0.75	21
Average monthly household consumption (KWh)	55	60	54	62	13
Average monthly household payment (MDL)	5 3	40	39	48	37
Average percent of income spent on electricity	5.3	5.3	4.2	3.9	−20
NRED					
Average tariff (MDL)	0.56	0.59	0.64	0.67	20
Average monthly household consumption (KWh)	52	52	56	60	15
Average monthly household payment (MDL)	29	31	36	40	38
Average percent of income spent on electricity	5.7	4.9	4.5	3.6	−33

Source: Tariff data are from ANRE 2002 and 2003. Consumption data are from the Moldova Household Budget Survey.

The study analyzed the factors contributing to differences in consumption using a multivariate model.[30] Those differences were more closely linked to location and income than to the electricity provider. Electricity consumption increased in cities and small towns served by Union Fenosa and decreased in cities and small towns served by the NREDs (table 6.4). The most significant difference was between Chisinau, served by Union Fenosa, and Balti, served by an NRED: consumption in Chisinau rose 16 percent, while consumption in Balti decreased by 13 percent. The

Table 6.4. Change in Electricity Consumption between 2000 and 2003, by Type of Provider and Location
(percent)

Item	Large cities	Small towns	Rural areas
Average household electricity consumption			
Union Fenosa	16	8	21
NREDs	−13	−3	31
Average share of income on electricity			
Union Fenosa	−24	−32	−31
NREDs	−39	−39	−36

Source: Moldova Household Budget Survey.

change was driven by faster income growth in Chisinau. Household electricity consumption rose 31 percent in rural areas served by the NREDs, and 21 percent in rural areas served by Union Fenosa.

The quality of service provided by Union Fenosa and the NREDs was also roughly similar. Service interruptions at Union Fenosa reportedly fell from 5,645 hours in 1997 to 52 hours in 2002.[31] Equivalent data were not available for the NREDs, nor were data available for such other measures of service quality as customer complaints and billing flexibility and accuracy. However, focus group discussions and interviews with consumers served by both companies suggest that the number of interruptions and voltage oscillations were similar.

Overall, electricity sales rose 47 percent between 1999 and 2002 (table 6.5). The increase in sales by Union Fenosa and the NREDs was similar, with increases of 49 percent and 45 percent respectively. Debts for energy imports steadily declined for private companies and increased for public ones, suggesting better performance by the private sector.

Power losses—consisting of electricity transmitted by distribution companies minus residential and nonresidential metered consumption—remained high for both Union Fenosa and the NREDs, imposing a significant cost on the sector. Between 1999 and 2002, commercial and technical losses decreased slightly from 31 percent of total power to 29 percent (table 6.6).[32] Union Fenosa losses fell between those of the two state-owned companies. Virtually all commercial losses were attributable to theft. Consumers were afraid of the large fines for theft, which could reach MDL 10,000, and this appears to have had a significant effect on performance. Union Fenosa reported a 9.3 percent fall in

Table 6.5. Net Sales at State-Run Electric Utilities and Union Fenosa, 1999–2002
(millions of MDL)

Company	1999	2000	2001	2002	Percentage Change between 2000 and 2002
NREDs	292	318	342	422	45
RED Nord	190	224	232	282	48
RED Nord-Vest	102	93	110	141	37
Union Fenosa	735	899	1,043	1,092	49
RED Chisinau	491	612	691	744	52
RED Centru	147	168	208	205	40
RED Sud	97	119	145	142	47
Total	**1,027**	**1,216**	**1,385**	**1,515**	**48**

Source: ANRE 2002 and 2003.

Table 6.6. Electricity Losses by Union Fenosa and the NREDs, 1999–2002

Company	1999	2000	2001	2002
RED Nord				
Volume of losses (millions of KWh)	251	155	155	138
Percent of revenues	38	28	28	24
RED Nord-Vest				
Volume of losses (millions of KWh)	133	98	120	115
Percent of revenues	36	36	40	35
Union Fenosa				
Volume of losses (millions of KWh)	678	722	753	655
Percent of revenues	28	32	34	29
Total				
Volume of losses (millions of KWh)	1,061	974	1,028	908
Percent of revenues	31	32	34	29

Source: ANRE 2002 and 2003.

theft,[33] attributing the decline to a new program to install tamper-proof meters.[34] Focus groups and interviews indicate that enforcement improved as meter readers were rotated more frequently and given a share of the fines they collected as a commission.

Quantitative data on theft were not available, but the qualitative findings suggest that it is not related to income. Indeed, it helps to have means or connections to steal: focus groups and interviews indicate that wealthy customers were most likely to steal. Theft was limited to those with the means to bribe meter readers, invest in technology to circumvent the meter, or steal using other means; people who could afford theft devices or had the technical skills to set up an illegal hook-up; small, energy-intensive enterprises, which often steal from other consumers; and poor households, which often steal only occasionally or because they were disconnected for failure to pay their bills.

How Effective Was the Social Transfer System?

In common with many other former Soviet Union countries, Moldova's current strategy for mitigating the impact of tariff increases, the nominative targeted compensation (NTC) system, is not targeted at the poor. Instead of being means tested, it is a system of categorical privileges: certain groups of people receive the NTC (box 6.2), which helps cover the cost of electricity, gas, district heating, hot water, cold water, coal, and firewood.

Even following reform of the system, which reduced the number of categories from 37 to 11,[35] the correlation with poverty is weak: the

Box 6.2

Nominative Targeted Compensation Categories

In accordance with Government Decision No. 761, as of July 31, 2000, compensation is paid to the following categories of people:

1. Disabled people belonging to groups I and II, regardless of the reason for their disability.
2. Disabled people belonging to group III who are
    a. Labor veterans.
    b. Recognized as disabled as a result of severe injuries, traumas, or wounds received during execution of military duties.
    c. Participants in military actions for defending the integrity and independence of the Republic of Moldova.
    d. Victims of political repressions between 1917 and 1990.
    e. Former prisoners of concentration camps or ghettoes.
3. Disabled children under age 16.
4. People disabled from childhood.
5. Participants in World War II and their spouses, depending on circumstances.
6. People whose status is equal to that of World War II veterans.
7. Parents, spouses who do not remarry, and the preadolescent children of people who were lost executing service duties or who died as a result of participation in attempts to control the accident at the Chernobyl Atomic Power Station.
8. Single pensioners.
9. Families with four or more children under age 18.
10. People who supported the troops during World War II.
11. People who were in Leningrad during its blockade.

Source: Moldova Ministry of Labor and the Social Protection (2003).
Note: The NTC is the Moldovan government's primary instrument for delivery of social benefit assistance.

proportion of households in the lowest 20 percent receiving the NTC was only slightly higher than for the highest 20 percent, 16 percent versus 14 percent (table 6.7). Moreover, the lowest 20 percent of households received the smallest share of NTC resources, while the highest 20 percent received the largest.[36]

Table 6.7. Households Receiving Nominative Targeted Compensation for Electricity, by Income Quintile

Quintile	Percent of households receiving compensation
Lowest	16
Mid–low income	17
Middle	15
Mid–high income	13
Highest	14

Source: Moldova Household Budget Survey.

The timing of NTC transfers was also important for the very poor. It was not aligned with the electricity company billing cycle, which made it harder for very poor consumers to pay their bills on time. Union Fenosa offered consumers a financing mechanism for smoothing payments, but few customers took advantage of it.[37]

Proposing a Better Mitigating Strategy

Despite positive news on rising income and the closing of the electricity consumption gap between the poor and nonpoor, electricity consumption remained exceptionally low and inelastic, especially for the very poor. Between 1998 and 2003, consumers reduced consumption and paid more for the power they used (figure 6.4). This implies large potential consumer welfare losses associated with future tariff increases unless accompanied by further increases in income. It also implies that there may be room for substantial welfare gains through enabling households to better manage their electricity expenditures. This could be achieved by introducing prepayment swipe cards for meters to reduce both the cost and the anxiety associated with disconnections, or encouraging the poor to use more energy-efficient technologies for refrigeration and lighting by introducing vouchers or similar programs. The public sector could also help defray the cost of extending access to clean, inexpensive gas in small towns, where people must rely on electricity for heating. But to achieve this, a financially sustainable gas sector is needed first.

More could be done to target the very poor, such as reformulating the NTC to make it more of an income-based transfer and promoting the optional lifeline introduced by the NREDs in June 2002. Under that program, customers pay MDL 0.50 for the first 50 KWh and MDL 1.65 for every KWh over 50. Interestingly, only about 10 percent of households served by the NREDs used this program.[38] Given the very low consumption of poor people, it is unclear why more poor households have not elected to participate. One reason may be fear of the very high expenditures associated with exceeding the 50 KWh threshold. It is probably worth exploring whether a different tariff structure would encourage more poor people to participate without compromising utility finances.

Conclusion

Moldova's energy sector has brought the country out of the energy crisis of the late 1990s. Electricity supply has increased, payments have gone up, and the sector's performance has improved. Government expenditures on fuel and energy decreased from MDL 36.9 million in 1997 to just MDL 2.1 million in 2003.[39]

The study showed that the poor benefited more than the nonpoor from reform, having increased their consumption more than the nonpoor despite rising costs. Consumption and expenditure patterns of households served by the private operator are roughly similar to those served by the public utilities. While the share of electricity in income fell more for the NREDs, it was similar for all consumers and lower than at any time since 2000.

Privatization did not hurt the sector. The private company had a significant positive impact on the government budget, while service quality improved (electricity is now available 24 hours a day), and collection rates have risen to almost 100 percent across the country. Indeed, privatization might have improved performance by state-run companies, highlighting another institutional factor that could influence the success of reform efforts: the coexistence of private and public distribution companies. With elements in Moldova keen to see privatization discredited, the presence of a private operator put pressure on the NREDs to show that publicly run companies could produce results equal to or better than a private operator, and thus improve their performance. If true, this implies that the presence of a private operator in a chronically underperforming sector may have a significant positive spillover effect. There may be significant advantages to applying a partial privatization model in other underperforming sectors, such as district heating, water, and possibly gas.

Another possible spillover effect of privatization was the rapid remonetization of the economy. Even before privatization, the barter system was beginning to disappear in Moldova, just as it did in the Russian Federation and Ukraine. It is likely that by refusing to engage in this highly inefficient but widespread method of payment, Union Fenosa hastened its demise in Moldova.

Notes

This chapter is based on World Bank 2004f.

1. In 2003, Moldova's per capita gross domestic product (GDP) was US$543, among the lowest in the region (National Bank of Moldova at www.bnm.md/english/index_en.html). Multiple years.

2. World Bank (1996b).

3. World Bank (2004c).

4. Together with Transnistria, Moldova imports 30 percent of its electricity, with the remainder produced by Moldova GRES in southern Transnistria from gas and oil purchased from the Russian Federation and Ukraine.

5. World Bank (2003e).

6. Dodonu (1999).

7. International Energy Agency (2003).

8. Moldova unbundled the state energy company into 16 generation, transmission, distribution, and debt-holding entities. In 1998, an electricity law was passed, and in preparation for privatization, an independent regulator, the National Energy Regulatory Agency (ANRE), was established to regulate gas and electricity (Electricity Law No. 137-XIV of 1998, cited in World Bank 2002b). In addition, debt was restructured and transferred to oldtranselectro, a state-owned debt-holding company (World Bank 2002b).

9. ANRE, multiple years.

10. Ministry of Energy. Data for the NREDs have not been made available.

11. These figures do not reflect the full economic costs of the different types of energy, which may include transport, storage, and costs to health. These costs may apply much less to utilities than to nonnetwork energy sources.

12. Of 36 urban centers that once had district heating, only 6 (including Chisinau and Balti) still had functioning systems, and service was not reliable.

13. These figures are based on household-level data collected from the HBS and Union Fenosa database. Figure 6.2 is different because its data are sourced from the International Energy Agency and represent aggregate consumption (including residential and nonresidential) divided by total population.

14. Derived from U.S. Department of Energy (1997).

15. The price of wood was MDL 220 per cubic meter in 2003, or about MDL 0.0012 kg oil equivalent. Wood use—which averaged about 0.5–1.5 cubic meters in an average summer month and 2–3 cubic meters in an average winter

month—cost households MDL 110–330 in the summer and MDL 440–660 in the winter. These figures are higher than those for gas: central gas expenditures were MDL 101 a month, and LPG expenditures averaged MDL 132 a month.

16. Interview with Deputy Energy Minister Felix Varlan.

17. Moldova Household Budget Survey, multiple years.

18. Union Fenosa data revcal no statistically significant difference at 1 percent level between payment (or collection) rates for the poor and the nonpoor.

19. There is no reason to believe that marginal increases in monthly KWh consumption will decrease. Average household electricity consumption in Moldova is far below that of its neighbors.

20. Data are not available for voltage and frequency fluctuations, which are a function of both generators and distributors.

21. A World Bank Poverty Assessment in 2003 found that large cities (Chisinau and Balti) had the lowest poverty rates, at 25 percent; poverty was higher in rural areas (38 percent) and highest in small towns (52 percent) (World Bank 2004c).

22. Calculating the household energy bundle would show how it changed over time with the change in the relative cost of fuels. Doing so is not possible using HBS data, however. Data on wood, coal, and other fuels are unreliable because of the small number of observations. Data on district heating expenditures are not believed to be reliable because many households apparently did not pay for this service.

23. Between 2000 and 2003, GDP rose by 21.6 percent and wages by more than 70 percent, and unemployment fell (Economist Intelligence Unit 2004).

24. Derived from data from TACIS Moldova Economic Trends.

25. ANRE (2003).

26. Interview with ANRE director Nicolae Triboi, June 15, 2004.

27. Union Fenosa's reconnection fee after debt payment varies by customer type, distribution company, distance, and other factors. RED Chisinau charges MDL 92, RED Centru MDL 14–62, and RED Sud MDL 45–57 to reconnect residential customers. RED Chisinau charges nonresidential customers MDL 201, RED Centru charges MDL 155–234, and RED Sud charges 80–180 (Union Fenosa data).

28. Data cover only the period after electricity distribution was split into private and state-owned enterprises.

29. The impact of tariff changes on the two networks differs. Households served by the NREDs had a lower elasticity than those served by Union Fenosa.

30. See World Bank (2004f) for details.

31. Union Fenosa data.

32. A concession for commercial and technical losses is included in the tariff methodology (17.7 percent for Union Fenosa, 18.0 percent for the NREDs in 2002), above which the cost is borne by the company (ANRE data).

33. Union Fenosa data comparing first quarter 2004 with first quarter 2003.

34. By March 2004, 133,749 new meters had been installed (Union Fenosa data).

35. The change was effected by the Law on Special Social Protection of Some Categories of the Population No. 933 XIV, passed April 14, 2000.

36. Counterpart International Study, based on a different data set and showing that compensation is poorly targeted.

37. The reasons for low participation were not found in the study. The study did find that for some households the NTC for electricity was higher than actual electricity expenditures. Some 15–20 percent of households receiving compensation for electricity were using the money for expenditures other than electricity.

38. ANRE data.

39. Moldova Economic Trends (2003). Data that could show the impact on the quasi-fiscal deficit—and thus quantify the impact of turning a debt-laden public entity into a tax-paying private company—are not available. In the future, it would be desirable for the government to maintain records to quantify the fiscal impact of privatization reforms. Neither the government, the International Monetary Fund (IMF), nor the World Bank were able to furnish these records.

CHAPTER 7

Timing and Sequencing of Raising Rates—Azerbaijan

The previous studies have given a better understanding of how tariff increases affect consumers, particularly the poor. The impact is most noticeable when tariff increases are sudden, leading to a dramatic decline in collections in Armenia and widespread opposition in Georgia. This study provides an ex ante analysis of an alternative approach, a more gradual increase in tariffs. In doing so it provides an idea of the advantages from reforming more gradually in cases where this is possible, and can offer reluctant governments empirical information on the consequences of alternative reform strategies.

Energy Rich, with Unrealized Power

Azerbaijan is a net energy exporter, a characteristic that radically alters the context of reform. Although it also experienced economic collapse and devastating conflict after the fall of the Soviet Union, it has not accumulated energy-related debts. And its natural resource endowment makes it less dependent on external assistance—so the government has more freedom to reject politically difficult reforms.

Despite being energy rich, Azerbaijan suffers from an unreliable domestic power supply. Power outside Baku is supplied a limited number of hours per day because of badly maintained infrastructure, high commercial losses, high nonpayment rates, and low tariffs. These problems are getting worse as strong economic growth increases demand for electricity. The opportunity cost of supplying the sector with low-cost domestic oil and gas is rising as international oil prices increase, and the government is sacrificing energy revenues.

To improve supply and reduce subsidies to the sector, Azerbaijan started energy sector reforms fairly recently. A key part of the reforms is raising tariffs to cost-recovery levels; at manat 96 ($0.0196) per kilowatt hour (KWh), residential tariffs are well below other countries in the region (table 7.1). Azerbaijan may be able to afford lower tariffs, but it must raise prices to cover generation, transmission, and distribution costs for the network to be financially viable. Without increased tariffs, the network will continue to decline, demand will outpace supply, and service quality will fall. When reforms were being considered, international norms suggested that cost recovery would be approximately manat 288 ($0.06) per KWh, an increase of 200 percent.[1]

Poor collection rates have further compounded problems associated with low tariffs. Collections from metered households in Baku (71 percent) are lower than in neighboring countries (see table 7.1).[2] Low collections reduce the tariff by half or more, the result of poor service quality, weak enforcement, theft, lack of metering, and nonpayment. Enforcement in Baku has improved in the last few years because of the presence of a private operator, Barmek, and collection rates are predicted to rise to 100 percent by 2008.

Table 7.1. Tariffs Are Lower and Consumption Is Higher in Azerbaijan

Country	Tariff (dollars per KWh)	Collection rate (percent of payment per billing)	Mean household consumption (KWh a month)
Azerbaijan (Baku, 2002)	0.0196	71	198
Moldova (Chisinau, 2003)	0.0529	98	58[a]
Georgia (Tbilisi, 2002)	0.0564	90	158
Armenia (Yerevan, 1999)	0.0475	82	169[b]

Source: See annex 4.
Note: Figures for Baku are based on records for 1,094 metered households in the 2002 Household Budget Survey.
a. January–November.
b. January–June.

Reluctant to implement politically difficult reforms, the Azerbaijani government expressed concerns about the social impact of increasing electricity prices, particularly the tariff level increase required for cost recovery. This study was to provide the government with information on the potential impact of tariff increases and the mitigating strategies to avoid welfare losses for consumers. To increase tariffs to cost-recovery levels would involve a tariff increase of 200 percent. Because of the government's concerns, the study simulated the impacts of an increase to cost-recovery levels and of smaller increases to show how different options might affect household welfare and sector sustainability.

Box 7.1

Data for the Analysis—Azerbaijan

The study began with a stakeholder analysis to identify elements of the reform package that were not supported by the stakeholders and why, using focus groups and interviews with key informants.[a] These were followed by a household budget survey (HBS) and quantitative analysis of the data to simulate the effects of various tariff increases on household consumption.

The welfare effects were measured as the amount of compensation the household would need to achieve the same welfare level as before the increase. The effects can be evaluated using results of an electricity demand model. The empirical strategy used in this study estimated the pooled model of electricity demand using household survey data for four countries: Armenia, Azerbaijan, Georgia, and Moldova.[b] These household survey data were merged, household by household, with the payment and billing records provided by the electric utilities for limited samples of households in the capital cities of each country. Pooling creates a data set with sufficient price variation to enable estimating the price elasticity of demand.

Estimation of a single model on the pooled data set assumes that the four countries have similar conditions, particularly in the household energy sector, a reasonable assumption since the countries share many common characteristics and are at approximately the same stage of transition. The biggest differences are in per capita income and access to substitutes, both accounted for in the model.

a. These included representatives from the Presidential Administration, Cabinet of Ministers, Ministry of Economic Development, Ministry of Fuel and Energy, Ministry of Labor, Ministry of Environment and Natural Resources, Parliament, Energy Sector enterprises, the media, and nongovernmental organizations.
b. The majority of observations in the data set came from Georgia.

Residential Energy Consumption

Average electricity consumption in Baku was well above basic minimum needs and higher than in other countries with similar data (see table 7.1). These findings were expected because of Azerbaijan's lower prices and collections. Consumption was not significantly higher, possibly because many households in Azerbaijan, particularly in Baku, had access to a reasonably reliable supply of inexpensive natural gas. Average electricity consumption for metered households in Baku was anywhere from 2,376 KWh[3] to 2,952 KWh per year,[4] or 198 KWh to 246 KWh per month.[5]

Metered households in Baku spent about 2 percent of their income on electricity in 2002 (table 7.2). This level of spending was similar to households in the United States (2.3 percent), but was well below those in the United Kingdom (4 percent) and most of the transition economies (generally 4–6 percent).[6] The low shares of income spent on electricity suggest there may be room to raise tariffs in Baku without severely limiting consumption of other goods and services.

There was little difference in consumption patterns between the poor and the nonpoor; in most countries, the lowest 20 percent of households spent a larger share of income on electricity than the highest. One explanation is that collections in Azerbaijan were lower for the poor (table 7.2), meaning that they faced a lower effective tariff than the nonpoor and consumed proportionally more than if collections were fully enforced.[7]

Reliable data on household electricity consumption outside Baku are not available because of lack of metering, frequent service interruptions, and high nonpayment rates. Average household consumption outside Baku and the Northeast—based on household data provided by the privatized

Table 7.2. Differences between the Poor and Nonpoor in Baku Are Small, 2002

Quintiles (per capita)	Household income (US dollars a month)	Household consumption (KWh a month)	Share of income on electricity (percent)	Collection rate (percent of payment per bill)
1 (poorest 20 percent)	123	190	2.1	65
2	137	202	1.9	61
3	154	192	1.9	74
4	161	201	1.9	68
5 (richest 20 percent)	189	200	2.2	81
Total	**158**	**198**	**2.0**	**71**

Sources: 2002 HBS, 2002 Barmek Records.
Note: Figures are based on records for 1,094 metered households in the HBS.

distribution company, Bayva[8]—ranged from 960 KWh a month in Imishly to 260 KWh a month in Mingecevir (table 7.3). The reliability of these figures is, however, highly questionable. For example, households with more hours of supply are expected to consume more, but the data show the reverse. One explanation is that households outside Baku were billed based on norms, so these figures represent expected, not actual, consumption. True electricity consumption outside Baku is not known. If electricity consumption outside Baku was as high as the data suggest, there may be opportunities to substantially increase the efficiency of electricity use.

It is not known whether electricity supply was rationed outside Baku, especially during the winter. No data are available from the utilities on the number of hours of electricity delivered to different locations. But households in Baku and Sumgait reported that electricity was available 24 hours a day. In other areas, supply was worse in winter (16 hours a day) than in summer (21 hours a day).[9] This is attributable to difficulties in supplying higher loads associated with residential consumption of electricity for heating. The results on hours of service are internally consistent—in different locations the majority of households reported similar hours of service. For example, all 150 households interviewed in Sumgait reported 24 hours of service. The results are also consistent with other surveys undertaken in Azerbaijan, suggesting that poor service outside the capital is a major impediment to economic development.[10]

Table 7.3. Electricity Consumption and Service Quality Vary Widely by Location

Location	Billing method	Mean household consumption (KWh per month)	Winter supply (hours per day)	Summer supply (hours per day)	Collection rate (percent of payments per billing)
Alibayramly	Norms	628	17	22	25
Baku	Meters	265	24	24	63
Ganja	Norms	na	10	22	na
Goycay	Norms	503	15	18	42
Guba	Norms	na	9	15	na
Imishly	Norms	960	8	20	7
Ismailly	Norms	na	18	21	na
Mingecev	Norms	260	9	21	28
Sabirabad	Norms	447	8	20	35
Sumgait	Meters	374	24	24	24

Source: 2003 Energy Survey (nonrandom) merged by household with 2003 Barmek and Bayva data (n = 2,000).
na = not available.

How Will Households Respond to Tariff Increases?

This section examines how a tariff increase would affect household electricity consumption in Baku, where there is no rationing constraint.[11] It then goes on to calculate the size of the income loss from different potential tariff increases—10 percent, 50 percent, and 200 percent—keeping everything else constant. It concludes by identifying who will be most affected by the tariff increase and what potential mitigating actions might imply.

Effect of Reform on Consumption

Understanding household responses to tariff increases requires knowing how much they reduce consumption in response to changes in price. To do this, a sensitivity analysis was first conducted looking at how consumption would change under a range of elasticities (low = –0.15, medium = –0.50, and high = –0.75). This was an informed estimate based on prior experience in the region. The impact of alternative tariff scenarios on household consumption was simulated based on these elasticities. The results show that large tariff increases combined with high elasticities cause dramatic falls in consumption (table 7.4).

The assumption of high elasticity is unrealistic, since, as seen in previous country studies, demand is likely to become more inelastic (less sensitive to price changes) as consumption approaches basic minimum needs. Even if electricity tariffs increased by 200 percent, it is unlikely that consumers would stop using electricity altogether. Also, the price elasticity of demand may change over time, and it is important to differentiate between short-term and long-term price elasticities. In the short run, elasticity is likely to be closer to zero than in the long run because a house-

Table 7.4. Changes in Consumption under Different Elasticities in Baku

Tariff elasticity	Consumption at current tariff level (manat 96)	Predicted consumption at 50 percent tariff increase (manat 144)	Predicted consumption at 100 percent tariff increase (manat 192)	Predicted consumption at 200 percent tariff increase (manat 288)
–0.15	200	185[a]	170	140
–0.50	200	150	100	na
–0.75	200	125	50	na

Source: Authors' calculations based on average consumption of 200 KWh a month.
na = not applicable because the value is negative.
Note: Collection rates are held constant.
a. Illustratute calculation: 185 KWh = 200 KWh – (0.50 x 0.15 x 200 KWh).

hold is better able to adjust to new relative prices of fuels and switch to cheaper electricity substitutes over a longer time period.

A realistic short-run scenario is that with a 200 percent tariff increase the elasticity is low; an informed estimate would be −0.15. At this elasticity, a 200 percent tariff increase will result in a fall in consumption from 200 KWh to about 140 KWh a month, a drop of 30 percent. So, all else equal, this analysis suggests that increasing the tariff by 200 percent to full cost-recovery levels would cause metered consumption of electricity for households with a 24-hour supply of electricity to fall to close to basic minimum needs.

Household Electricity Demand Model

This sensitivity analysis shows how the impact of a tariff increase depends on the price elasticity of demand. To produce a more reliable assessment of how household consumption and welfare will respond to price changes, the study created a household electricity demand model. The model was estimated by pooling household data sets and utility billing and payment records from capital cities of Armenia, Azerbaijan, Georgia, and Moldova.

In the model, which applies to urban households with meters, household electricity consumption depends on the tariff, household income, the household's access to substitute energy sources (natural gas, central heating, or liquefied petroleum gas [LPG]) and other household characteristics. Other important factors include location, daily temperature, and cross-country differences, such as economic growth and inflation. The model was estimated using multivariate regression techniques. With this type of modeling exercise the results are usually more reliable for small price changes than for large changes.[12] The model fits the data well and produces plausible results, providing a reasonably reliable basis on which to estimate the impact of tariff increases in Azerbaijan.

According to the model, a 10 percent increase in the price of electricity results in a 2 percent decrease in household electricity consumption—making the price elasticity of energy demand −0.20. This is very close to −0.15, the lower range of the sensitivity analysis presented earlier, and is also reasonably consistent with the studies that have estimated residential electricity demand in other parts of the world.[13]

Consistent with expectations, the model indicates use of central gas and LPG are negatively correlated with electricity consumption. Also as expected, increasing the collection rate (more enforcement) was negatively correlated with consumption. The model can predict changes in consumption under different tariff scenarios based on the -0.20 elasticity.

This model provides the income elasticity of electricity consumption. It indicates that a 10 percent increase in income will produce a 1.2 percent increase in consumption of electricity, so the income elasticity of electricity consumption is 0.12. The significance of this finding is clear: future household income growth will help offset the blow of a tariff increase, and incomes in Azerbaijan are expected to grow rapidly in the next few years; an increase in the minimum wage is being contemplated and civil servant wages were recently increased 50 percent. Therefore, calculating the negative impact of tariff increases on consumption and welfare levels without taking into account the positive impact from changes in income is the worst-case scenario.

Assuming current income of US$158 per household a month and a price elasticity of demand of -0.20, under a variety of tariff scenarios—in this case, increases of 50 percent, 100 percent, and 200 percent—income growth of 10 percent will keep the share of income for electricity around 4 percent (table 7.5). Depending on how quickly incomes grow and, more important, how growth is distributed between the poor and nonpoor, this will bring shares of income for electricity in Azerbaijan closer to the level in other transition countries. Surprisingly, model testing revealed no plausible significant differences in the price and income elasticity of demand for the poor and nonpoor.

Table 7.5. Rising Income Will Offset the Blow of Tariff Increases on Baku Households' Budget Shares
(percent)

	Share of income on electricity		
Tariff	Household income growth at 0 percent	Household income growth at 5 percent	Household income growth at 10 percent
Current tariff (manat 96)	2.5[a]	2.4	2.3
50 percent increase (manat 144)	3.3	3.2	3.0
100 percent increase (manat 192)	4.0	3.8	3.6
200 percent increase (manat 288)	4.5	4.3	4.1

Source: Authors' calculations based on average consumption of 200 KWh a month and price elasticity of demand of -0.20.
a. Illustrative calculation: 2.5 percent = (200 KWh x $0.02) / $158 a month.

The empirical data on the impact of different tariff increases on household electricity consumption are an important input into policy making because they offer a reliable measure of how much worse off households will be if different policy options are taken. They also suggest that small, gradual tariff increases rather than abrupt, large ones will soften the blow to household income, since this will allow time for income growth to offset the increase in electricity prices.

How Much Households Need to Be Compensated

The study also calculated the income loss from a tariff change using linear approximation. The maximum, or upper bound, of this loss is the additional amount of money that the consumer would have to pay after the tariff increase if electricity consumption is held constant. This assumes zero price elasticity of demand. The minimum, or lower bound, is the additional amount of money that the consumer would have to pay at the new tariff if their electricity consumption falls in response to higher prices, assuming the price elasticity of −0.2 calculated in the demand model.

If consumption before the tariff increase was 200 KWh—and assuming 100 percent collections—then the upper bound on the income loss from a 50 percent tariff increase would be manat 9,600 (US$1.95) per month, and the lower bound manat 8,640 (US$1.76).[14] So, the average welfare loss in dollar terms from a 50 percent tariff increase in Baku would be close to US$2 per household per month. This is the amount of money that would have to be given to a household to make it no worse off than it was before the tariff increase. The study made this calculation under various tariff scenarios, including a 200 percent increase to cost-recovery levels (table 7.6).

Table 7.6. Household Consumption and Income Loss under Alternative Tariff Scenarios
(elasticity is −0.2)

Percent increase in tariff	Tariff (manat)	Tariff (dollars)	Electricity consumption (KWh)	Maximum income loss (dollars per month)	Minimum income loss (dollars per month)
0	96	0.02	200	0	0
50	144	0.03	180	2.0	1.8
100	192	0.04	160	3.9	3.1
150	240	0.05	140	5.9	4.1
200	288	0.06	120	7.8	4.7

Note: Income loss calculated to one decimal place. Authors' calculations.

Differences between the Poor and Nonpoor

Calculating the effect of a tariff increase is complicated by lower collections for the poorest 20 percent of households than for the richest. This means that the poor are more vulnerable than the nonpoor when rising collections are taken into account; they face a bigger effective tariff increase than the nonpoor if collections are uniformly enforced. This implies that the poor require slightly more compensation than the nonpoor if tariffs and collections increase simultaneously. For example, a nominal 50 percent tariff increase to manat 144 per KWh, and enforcement of this tariff, will result in a higher effective increase for lower quintiles than for the higher quintiles. In this situation, to maintain constant welfare levels, the poor require around US$3 a month, whereas the nonpoor require closer to US$2.50 a month (table 7.7).

As in the previous studies, understanding who accumulates arrears has important implications for the welfare effect of reforms. Because mainly the poor accumulate arrears in Azerbaijan, affordability is a problem and special care must be taken by the state to provide adequate assistance to them.

Availability of Substitutes

Households without access to gas or wood in the capital towns of Rayons and rural areas may be particularly vulnerable to tariff increases. There are no good substitutes for electricity for lighting, refrigeration, and television. However, wood, kerosene, LPG, and gas, if available, are viable substitutes for electricity in heating and cooking. Households that do not have access to these alternatives will have the greatest difficulty in

Table 7.7. Compensation for the Poor in Baku Should Be Higher

Welfare quintiles	Current effective tariff[a]	Current consumption (KWh) per month)	Tariff tariff[b] (manat)	Predicted consumption (KWh) per month)	Minimum loss (dollars per month)	Maximum loss (dollars per month)
1 (poorest)	62	190	82	140	2.3	3.2
2	59	202	85	144	2.5	3.5
3	71	192	73	153	2.3	2.9
4	65	201	79	152	2.5	3.2
5 (richest)	78	200	66	166	2.2	2.7

Source: Authors' calculations assuming elasticity of –0.20.
a. Collection rate x manat 96.
b. manat 144 – effective tariff.

shifting their energy consumption to less expensive fuels, making them more vulnerable to tariff increases.

Dividing households around the country into groups based on location and access to gas and wood revealed that typical electricity consumption was significantly higher (600–700 KWh per month) among households in "other urban" and "rural" areas that did not have access to gas or wood (table 7.8). In these areas, very high percentages of households reported heating only with electricity. These households will be particularly vulnerable to tariff increases, especially if there are no improvements in service quality.

How to Mitigate the Impact of Tariff Increases

Increase Tariffs Gradually

Future household income growth will help offset the burden of a tariff increase, particularly if tariffs increase gradually. A gradual increase will soften the blow to household income, since price elasticity is likely to be greater in the long run than in the short run.

Link Tariff Increases to Service Quality

Experience shows that opposition to tariff increases can be avoided by explicitly linking tariffs to improved service quality. Raising tariffs and enforcing disconnections is unpopular, and the public often views state actions in this sector with skepticism. Consumers are especially skeptical when tariffs increase without any improvement in the quality of service because the costs (higher tariffs) come before consumers see the

Table 7.8. Households with Less Access to Substitutes Consume More Electricity

Location	Access to gas	Access to wood	Electricity (KWh per month)	Households heating with electricity (percent)	Hours of winter supply
Baku	Yes	No	246	12	24
Other urban	Yes	No	403	19	16
Other urban	Yes	Yes	361	2	16
Other urban	No	No	713	76	12
Other urban	No	Yes	427	33	9
Rural	Yes	Yes	136	0	22
Rural	No	Yes	504	2	10
Rural	No	No	608	18	12

Source: 2003 Household Energy Survey.

gains (improved service). Qualitative evidence gained in focus groups confirmed that households were afraid that they would end up paying more and still not receive sufficient supply of electricity. Investments in rehabilitation and maintenance of the infrastructure will help generate popular support for the increases in enforcement and tariffs necessary to finance such investments, especially outside Baku where service quality is worse.

Improve Efficiency of Energy Use

Households with access to few alternatives to electricity should be given access to efficiency-increasing technology and appliances, less expensive fuels for cooking and heating, and household insulation.

Improve Access to Clean Substitutes

Another option, if provided on a full cost-recovery basis, would be to encourage use of such clean and inexpensive substitutes for heating and cooking as natural gas. In Baku, 33 percent of households heated with electricity, 12 percent only with electricity, and their average annual consumption was 3,363 KWh.[15] This was about 615 KWh per year more than households that do not heat with electricity. Unless they can start heating with gas, they will require an additional US$5–$6 a year in compensation for a 50 percent tariff increase. Access to substitutes can be provided through a variety of instruments, as long as the government explicitly compensates the utility for any social transfers it provides. For example, the government could bid out competitive subsidies to encourage the extension of natural gas networks to poor neighborhoods. While the households would still have to pay the full cost of gas, the cost of bringing the network to them could be partly financed by the public sector.

Consider Lifeline Tariffs or Direct Transfers

There is no easy answer when considering the tradeoffs between alternative social protection strategies. The government can mitigate the welfare effects of tariff increases by providing assistance to poor and vulnerable households and by stimulating income growth.[16] In deciding between lifeline tariffs and targeted cash transfers, the Azerbaijani government needs to consider such factors as the percentage of those living in poverty and the targeting effectiveness of social assistance schemes. Given the influence of location on poverty in Azerbaijan, a geographically targeted transfer or lifeline tariff could be a highly effective and easily implemented solution. Indeed, this solution would be worth further investigation by the Azerbaijani government.

Outside Baku

Clearly, better data are required on electricity consumption and substitution behavior outside Baku before a definitive conclusion can be drawn on the magnitude of the impact. One solution would be to pilot metering where households have little access to substitutes to observe actual electricity consumption. These data could then be used to determine the best mitigation strategy for such households.

Conclusion

Energy sector reform is highly sensitive for the Azerbaijani government, which fears opposition to tariff increases, particularly at critical times in the presidential election cycle (the last presidential election was in late 2003). It is also a sensitive topic for the donor community, which must find a way to manage the government's opposition. As seen in chapter 6, misconceptions about the effects of reform can undermine the positive effects and threaten the sustainability of reform. And interviews with key informants within the government, business, media, and nongovernmental organizations revealed that though there was consensus on the need for reform (tariff reform, mitigating strategies, improved service, and private sector participation), many stakeholders felt poorly informed and raised concerns about reform. Nonenergy enterprises and the general population need to be informed about the potential benefits of reform. Nonenergy enterprises were concerned about losing competitiveness because of higher production cost from higher electricity tariffs. Households were afraid they would end up paying more and still not receive a sufficient supply of electricity.

In such a context, the advantages of ex ante analysis for designing reform measures and mitigating actions are clear. In this case, the study produces several useful policy prescriptions, based on an empirical simulation of the welfare impact of reforms on stakeholders. These ex ante insights arm the reforming government with powerful information on the full set of policy choices available.

The qualitative data also provide valuable insights on attitudes toward reform. Perhaps most influential is the level of skepticism over tariff increases and the promises of improved electricity supply. This insight underlines the importance of linking tariff increases with improvements in service quality to enhance confidence in reforms. The study can help a targeted public information campaign address consumers' concerns and build wide public support for the reform program.

More broadly, this study expands understanding of how tariff increases affect the poor. By using an electricity demand model it was able to predict the welfare impact of different tariff increases before they were implemented, providing a critical tool in evaluating different reform options. It suggests that small, gradual tariff increases are better than one large increase, because elasticity is greater in the long run and rising incomes help soften the blow.

Although future income growth will help offset effects of a tariff increase, for a government trying to balance sector sustainability with political sustainability of reform efforts, gradual tariff increases will be far easier for the adjustment of low-income households. People have more time to switch to substitutes and become better off as incomes rise. This approach offers an alternative to the view proposed in the early 1990s: that reform must be undertaken swiftly to be successful. It also comes with a significant caveat; countries poorer than Azerbaijan may not be able to wait before introducing cost recovery to their utilities. But for countries like Azerbaijan, which can afford to adjust more slowly, and where external advocates of reform enjoy a little less leverage, reliable information on the consequences of various options is a valuable input into policy debates.[17]

Notes

This chapter is based on World Bank 2004g.

1. This figure would be equal to the long-run marginal cost of a greenfield power plant in the United States. More careful country-specific calculations have shown that cost-recovery levels for Europe and Central Asia tend to be lower, highlighting the need for careful analysis of local conditions prior to reform.

2. Collections from metered households were significantly higher than general collection rates for urban and rural households, which were 50 percent and 30 percent respectively. 2003 Energy Survey (nonrandom) merged by household with 2003 Barmek and Bayva data ($n = 2,000$).

3. 2002 Household Budget Survey data merged by household with 2002 Barmek data ($n = 1,106$).

4. 2003 Energy Survey (nonrandom) merged by household with 2003 Barmek data ($n = 443$).

5. The lower figure is more reliable because it is based on a larger, more representative sample.

6. Moldova (5 percent), Georgia (5 percent), and Armenia (8 percent).

7. Not as much separation in the quintiles was observed as might be expected because income is presented on a household, not a per capita, basis.

8. The Barmek service area is Baku and the Northeast, and Bayva covers everything else. Barmek records are for the month of November 2003 only.

9. These averages are over the 2,000 households in the 2003 Household Energy Survey.

10. Foreign Investment Advisory Service (2002); World Bank (2003f).

11. The impact of relieving the rationing constraint—the positive effect of increasing electricity supply—cannot be assessed because the data on household behavior outside of Baku are not reliable. In Baku, consumption levels are based on actual consumption. The only data available on consumption in rationed areas are based on norms, not actual consumption.

12. A detailed description of the data and model is included in annex 4.

13. Because Azerbaijan started with a higher consumption level, consumption was initially more elastic. A change in tariff would result in a proportionately higher fall in consumption compared to other countries where the initial base consumption was lower. This also means that the estimate of compensation later on in the chapter is marginally higher than it would be otherwise.

14. Lower bound = manat 8,640 = (manat 144 – manat 96) × 180 KWh a month. Upper bound = manat 9,600 = (manat 144 – manat 96) × 200 KWh a month.

15. These figures come from the 2003 Energy Survey (see box 8.1).

16. See chapter 9 for discussion on this topic.

17. Since this study came out in late 2004, Azerbaijan has introduced a power sector reform project financed by the World Bank and has committed to bringing electricity tariffs to cost-recovery levels by 2010. The government has also implemented sharp tariff increases in gas and water.

CHAPTER 8

Coping with the Cold: Heating Strategies for the Urban Poor

A significant part of energy demand in Europe and Central Asia (ECA) is determined by a single characteristic: the region's extremely cold winters. Without reliable heat provision during winter, all aspects of everyday life are affected. Households rely on energy to generate warmth for survival. Businesses rely on it to operate. Without heat, such public institutions as schools and hospitals are forced to close or operate at close to freezing temperatures. Unless there is access to clean, affordable heating, the burden of heating expenditures becomes unsustainable, and households must resort to substitutes (wood and coal) that carry substantial negative environmental and health externalities. The social, economic, and political ramifications of inadequate heat supply make the responsibility of governments critical. They also make ECA the only region where the World Bank routinely lends for heating.[1]

With district heating systems deteriorating as power sector infrastructure collapsed, reform programs focusing on district heat have been an integral part of energy sector reform. But as with electricity sector reforms, a radically changing environment of reduced incomes and disappearing state subsidies makes it vital to understand household demand for heat and the impact of policies on the poor. Without this knowledge it is difficult to design investments that are appropriate and effective in the local context. Using household level data, this chapter builds a picture of demand for this basic component of energy use, and then makes recommendations on appropriate interventions.

Inefficient District Heating Systems

In the 1950s, large, centralized district heating became the system of choice in most developed countries, including Eastern Europe and Central Asia. It is generally considered the most comfortable, efficient, and environmentally friendly heating mode, particularly for densely populated areas. And it often has the potential of efficiently using the waste heat recovered from combined heat and power (CHP) plants. In ECA, most residences in urban areas were connected to the system, unmetered and for a nominal fee. Users had no influence over when and how much heat was provided, but could be reasonably sure that it would be provided, for free, as soon as outside temperatures dropped below 8° Celsius (C) for at least five days. Rooms would be heated to at least 20°C most of the time and, lacking individual controls, consumers would respond to overheating by opening windows.

Transition, rising energy prices, and economic collapse brought difficult choices to governments trying to rationalize their budgets. Years of neglect and lack of investment have led to inefficient district heating systems, deteriorating service quality, and badly needed repairs. Experience in restructuring Soviet-type district heating systems in Eastern Europe had shown that, through a combination of investments, institutional improvements, and sector reform, district heating systems could be modernized to approach efficiency, cost, and service levels of Western Europe.[2] In the 1990s, international financial institutions, including the World Bank and the European Bank for Reconstruction and Development, took an active role in funding rehabilitation investments for district heating in many cities in the region. As part of these donor-funded projects, many governments in ECA reduced general subsidies for heat and raised prices for district heating.

In making people pay for heating, however, household demand became an important consideration. As discussed in chapter 3, as prices increased and incomes fell, there was a significant contraction in demand for energy in the region. The solutions that worked elsewhere were not fully applicable when devising heating solutions for households in extremely poor countries, particularly in small, rural towns. Though district heating can be the most efficient system, lack of metering in old systems and high fixed costs made it very difficult for customers to control expenditures, which particularly hurt the poor. And the absence of meters and the technical and political difficulties of disconnecting nonpaying customers made it almost impossible to enforce payment. The net result was a low-level equilibrium trap where on one side, often due to political pressure, governments continued to pump money into antiquated and failing district heating systems. On the other side, consumers refused to pay their bills for a service that used to be very low cost or free and kept deteriorating, or was too expensive, or provided more heat than they demanded.

With increasing evidence that the prevailing practice of rehabilitating district heating may not always be adequate, this study set out to shed light on the demand for heat and to recommend new ways to provide the poor, particularly the urban poor, with access to clean, affordable heat. Studying how people heat themselves when left to their own devices provides insights into how much energy they demand for heating and how much they are willing to pay for it. It also provides important information on what fuels they use as substitutes and what issues need to be addressed.

Box 8.1

Methodology and Data Sources—Heat Demand

Determining household demand for heat is difficult using a household budget survey (HBS). It requires being able to separate the demand for heat from nonheat energy, which can be confusing because households consume a mix of fuels for a variety of purposes. One household may use wood for heating and cooking in the winter and LPG for cooking in the summer; another household may use electricity for heating and gas for cooking in the winter and electricity for air conditioning and gas for cooking in the summer. One approach to identifying heat consumption is to use norms to net out basic needs, then study what is left over of expenditure. But that approach obscures the variations in consumption and spending patterns that are of interest.

To get around these problems, a new approach for estimating heat demand was developed that exploits a natural experiment deriving from data collected in Armenia, the Kyrgyz Republic, and Moldova, where deterioration of district heating has meant that it is no longer available for households in some neighborhoods who must now use other means to heat themselves. The approach relies

(Continued)

on splitting the data into two subsamples. The first subsample consists of house-holds that are connected to the central heating network and report that central heating is their only source of heat. For this group, all noncentral heat energy consumption will be for such nonheating purposes as lighting and cooking.

The second subsample is households that have no central heat and must rely on other means for heat. Their energy consumption will include consumption for heat and nonheat purposes. Comparing the total energy consumption (not in-cluding central heat) of these two groups of households makes it possible to isolate the energy used for heating of the second group. The data used for the model is from a sample of urban households from Armenia, the Kyrgyz Republic, and Moldova from 1999.[a] The results can be seen in box figure B8.1.

Figure B8.1. Energy Consumption Scatterplots

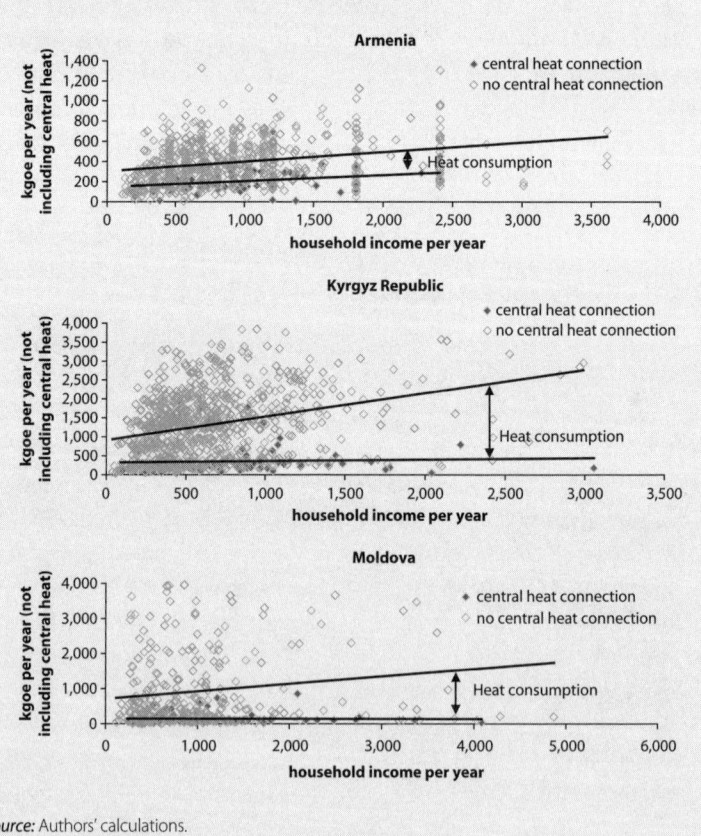

Source: Authors' calculations.
Note: kgoe = kilograms of oil equivalent.

(Continued)

The main disadvantage of this approach is that the demand for energy for heating is measured, rather than the demand for heat itself. The demand for heat cannot be measured directly because there are no data on indoor temperatures or the efficiency of heating appliances. This lack of data prevents directly exploring how much variation there is in actual heat consumption between the poor and nonpoor.

Household Demand for Heat

For households not on district heating networks, the poor are more likely to use traditional fuels such as wood (Armenia) and coal (Moldova), while the nonpoor rely on clean fuels such as electricity and central gas (figure 8.1).

These patterns have important implications for heating interventions. First, as incomes fall, people buy traditional heating fuels. Second, while cash transfers may offset the welfare effects of higher heating prices, they will not stop households from using traditional fuels if the prices of those fuels are not raised as well.[3] Thus, thought should be given to designing heating policies that take into account the social costs of burning traditional fuels. These include the health costs associated with not having

Figure 8.1. Urban Household Heating Fuel Choices by Income Quintile

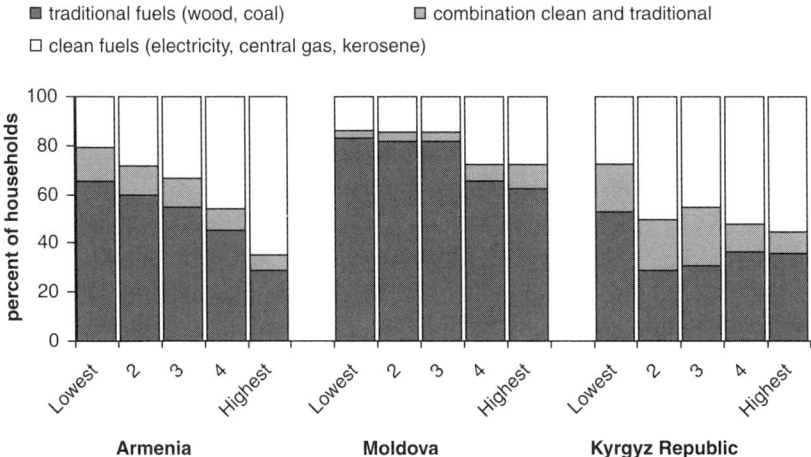

Source: Author's calculations from 1999 household survey data.
Note: Excludes district heating.

enough heat and the resulting productivity losses, the health costs associated with burning traditional fuels, the environmental costs associated with deforestation, and the opportunity costs of time spent collecting heating material, especially wood.

Estimating the Demand for Heat

The study estimated the income and price elasticity of demand for heat using a heat demand function. This was derived by plotting predicted heat consumption against price per kilogram of oil equivalent (kgoe) for three countries, Armenia, the Kyrgyz Republic, and Moldova. A heat demand function is expected to be kinked. It slopes steeply around the minimum amount needed for survival, and then rapidly levels off as the quantity of heat consumed goes from necessity to luxury. Identifying the location of this kink is important to understand how consumers respond to heat prices. At prices below the kink, demand is elastic and welfare losses resulting from a price increase are small, since households can still respond to the price rise by cutting consumption. At prices above the kink, demand is inelastic and welfare losses are large, because above this price households have already reduced consumption to basic minimum needs and cannot make do with less even as prices increase further.

A scatter plot of predicted household heat consumption against price per kgoe for Armenia, the Kyrgyz Republic, and Moldova suggests a function of precisely this shape (figure 8.2). There is a steep downward slope

Figure 8.2. Demand for Heat in Selected Countries

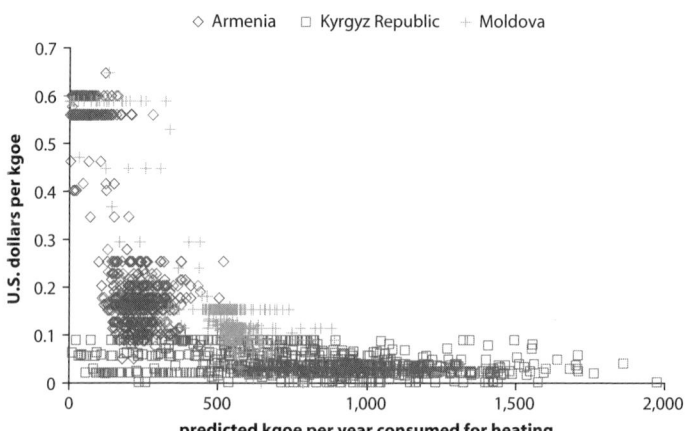

Source: Authors' calculations.
Note: Excludes district heating.

at prices above US$0.20 per kgoe (indicating the inelastic part of the demand function), followed by a rapid flattening out. It appears that households alter their heating strategies quickly in response to price changes in the range of US$0.01–$0.20 per kgoe, below which price demand is elastic. For households without substitution opportunities, welfare losses will be greater when the price rises above US$0.20 per kgoe, the inelastic part of the curve. In these cases it will be particularly important to design policies that cushion the blow of energy price increases on the poor.

This model suggests that the income elasticity of demand is between 0.1 and 0.2, meaning that a 10 percent increase (decrease) in income will produce a 1 percent increase (decrease) in energy consumption for heating by the poor, and about a 2 percent increase (decrease) by the nonpoor. As expected, demand is less elastic for the poor than for the nonpoor. That the three data sets produce similar results and are consistent with economic theory increases confidence in the model.[4]

As expected, there is much greater variation in price response by income group and country. Price elasticity is –0.4 in Armenia and –0.2 in the Kyrgyz Republic and Moldova, meaning that a 10 percent increase in price will produce about a 4 percent decrease in consumption in Armenia compared with about 2 percent in the Kyrgyz Republic and Moldova. In Armenia and Moldova, the poor are less price elastic than the nonpoor. That the poor are less income and price elastic than the nonpoor suggests that they will have greater welfare losses from price increases unless they can find less-expensive substitutes.

Although the elasticity and the point at which demand becomes inelastic will vary by country, this analysis provides policy guidance on the price above which consumer welfare begins to drop quickly and complementary interventions to address this drop may be needed.

Household Heat Consumption

Household heat consumption was estimated using the above model, and the results on a per capita basis are presented in figure 8.3.[5] The figure reveals variations in household heat consumption.[6] In Armenia and the Kyrgyz Republic, the poor consume less heat per capita than do the nonpoor.[7] The results are confounded by larger low-income household size, complicating the design of pro-poor heating tariffs such as lifelines, which are based on a minimum consumption level per household.

Annual nonheat energy consumption ranges from 50 kgoe per capita in Armenia to about 125 kgoe in the Kyrgyz Republic. Annual predicted

Figure 8.3. Predicted per Capita Heat and Nonheat Energy Consumption in Selected Countries

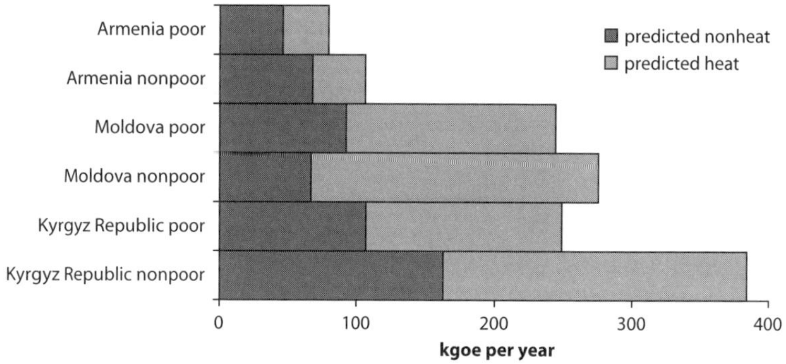

Source: Authors' calculations
Note: Excludes households on district heat.

heat consumption ranges from 40 kgoe per capita in Armenia to 175 kgoe in Moldova to 180 kgoe in the Kyrgyz Republic. Thus heat consumption accounts for 40–60 percent of total energy consumption. Differences across countries are driven by differences in climate and energy pricing policies. The average temperature during the heating season is highest in Armenia (2.6°C), followed by Moldova (0.6°C) and the Kyrgyz Republic (–2.9°C). Energy prices are highest in Armenia, followed closely by Moldova, and are substantially lower in the Kyrgyz Republic.

Household Heat Expenditure
To calculate heating expenditures, the study multiplied the predicted heat consumption by the price of a household's primary heating fuel, which was obtained from the survey. These calculations indicate that heating accounts for 5–10 percent of household spending and for 20–40 percent of energy spending. On average, the poor spend almost twice as much of their household budgets on heating as do the nonpoor (figure 8.4). In absolute terms, poor households spend US$25–$40 a year on heating and nonpoor households spend US$30–$50 a year.

These findings are important for three reasons. First, the fact that poor households spend a larger share of their budgets on heating suggests that it is possible to design a heating subsidy that benefits the poor more than the nonpoor. Second, that heat is a large share of energy spending suggests higher heating prices will considerably reduce household welfare unless inexpensive substitutes are available. Third, poor people are unlikely to

Figure 8.4. Predicted Heat Expenditure as a Percentage of Household Expenditures

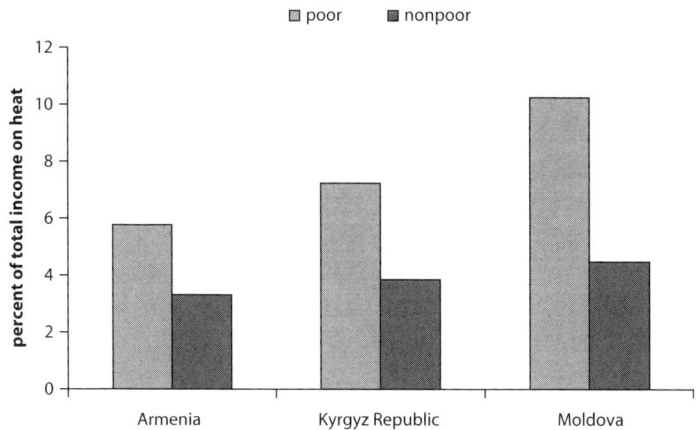

Source: Authors' calculations.
Note: Excludes households on district heat.

pay for heating systems that cost more than US$25–$40 per year because they can find less expensive ways to heat themselves (they might, however, be willing to pay slightly more for heating systems that are substantially more convenient).

Though we do not have data on actual heat consumption, the findings from this methodology for estimating heat demand are backed up by HBS data. In a survey, Armenian apartment dwellers were asked to estimate their previous year's spending on heating and their average indoor temperature during the heating season. Self-reported spending ranged from US$10–$20 a year, the same order of magnitude as the model results. Also, poor households with full control of their heating keep their apartments at lower temperatures and spend less than households on the district heating network. This finding backs up the central finding of our heat demand model: district heating designed based on a norm of 28°C provides more heat than consumers demand or are willing to pay for.

Rethinking Heat Supply

An understanding of heat demand is essential to designing suitable strategies for supplying heat. Before the transition, consumers connected to central heating in ECA expected that every room in their living quarters would be heated to about 20°C for 24 hours during the official heating season. Under such conditions, and with high population density, the heating system that provides heat at the lowest cost is district heating supplied from cogeneration plants.[8]

But many poor urban households consume less heat and have lower heat expenditures than usually associated with a district heating system. Lower household heat demand is manifested in lower supply temperatures, shorter heating seasons, and less area heated.

This section compares typical costs of various heat supply options for two levels of heat demand: full service, meaning provision of about 18°C in all rooms of a dwelling,[9] and reduced service, meaning a lower temperature in one or several rooms. Full service is the demand that is assumed when district heat is supplied; reduced service is a closer approximation of the actual demand revealed by the analysis above.

The heat supply options compared range from highly centralized district heating networks, fed by cogeneration plants or heat-only boilers, to building boilers that supply only one or a few buildings with heat, to decentralized (individual) heating where each dwelling has its own heat source. Each of these heating options can be based on a wide range of fuels and come with very different levels of efficiency and environmental performance. The costs of these options at the different levels are then compared with typical household expenditure levels. This yields conclusions about how to implement financially and environmentally sustainable and affordable heating strategies that take into account the fixed and variable costs and investment requirements of various heat supply options.

The Cost of Full Service

The costs of modernized district heating systems in countries and cities have been well researched during the preparation of feasibility studies. The resulting costs per unit of heat delivered at the building entrance usually fall within a fairly similar range of US$0.20–$0.35 per kgoe, leading to annual household heating bills of US$200–$900, depending on dwelling size, specific heat consumption, and heat tariff level.

How does this figure compare with the costs of other heating options for full service? Though these figures are less well known in the region, recent studies from Armenia suggest these options cost between US$135 and $324 a year for full heat service (figure 8.5). There is a large variation not only in the annual costs but in the capital (fixed) and fuel (variable) costs of different options, with natural gas having high investment costs and low fuel cost, while the opposite is true for heating based on electricity, kerosene, liquefied petroleum gas (LPG), and wood.[10]

Figure 8.5. Annual Costs of Different Heating Options for Full Heat Service in Yerevan, Armenia

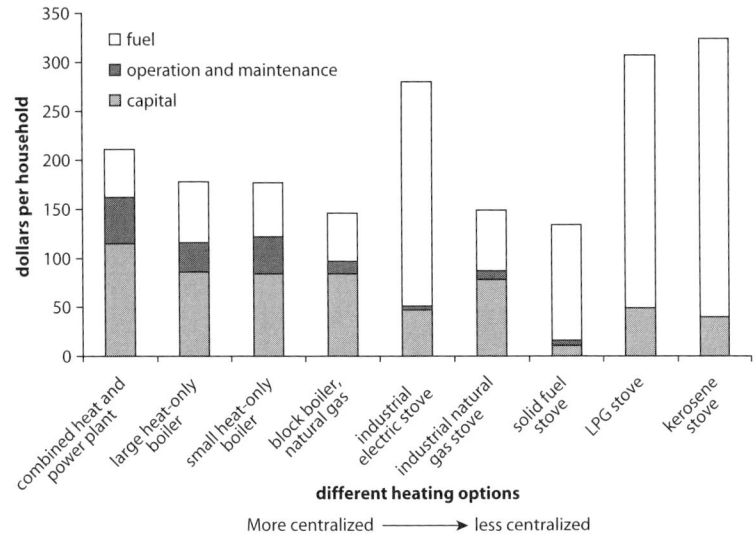

Source: Based on COWI (2002a).
Note: The calculations are based on a comfort level of 17°C and 110 heating days.

The Cost of Reduced Service

What are the costs of these different options for reduced service supply? Although district heating systems can be the most convenient and cost-effective heating mode given a heavy heat load, their high fixed costs make them expensive for consumers demanding less heat. Only for those households not on the network would reduced heat consumption result in lower heat bills. Those still connected to district heating experienced rising heat tariffs and higher expenditures despite declining service levels. This is because when heat supply companies lose customers, the old parts of their district heating systems do not permit heat not consumed in one place to materialize as fuel savings at the heat generation plant.

This characteristic means that utilities are typically not able to reduce costs in the short to medium term in proportion to the decline in demand. Typically, district heating systems can only be adapted to a lower heat load in the medium to long term with replacement investment and modernization of the system configuration. In the interim, the remaining customers have to bear even higher costs. In Bulgaria, this vicious circle could be observed in 1996–99. Since then, customers have slowly started to reconnect because of efforts to meter heat consumption and bill customers accordingly.

More flexible options such as individual heat technologies, for which fuel accounts for a larger share of total costs and which are modular, are much easier to adapt to the lower heat demand demonstrated and are more cost-effective with reduced demand. With electrical heating, for example, fuel accounts for about 85 percent of total costs (figure 8.6). Therefore, while electrical heating has a high unit cost, it may be less expensive for the household to heat with because it is more flexible.[11]

This suggests that district heating is inefficient and inappropriate for meeting new heating demand patterns. Centralized options are cheaper than electric heating or wood stoves when providing full heat service, but individual options are less expensive than centralized options for reduced service because they tend to be modular (figure 8.7).

In some cases, district heating may remain the most appropriate option. There are compelling factors favoring maintenance of carefully planned and affordable district heating systems in countries with relatively modern CHP plants that are needed for the power system, such as Moldova and the Kyrgyz Republic.[12] In densely built urban environments, individual heating is usually more expensive than any form of central heating at full service levels and can have negative environmental impacts, including

Figure 8.6. Fuel Costs as a Share of Total Heat Costs for Different Heat Supply Options and Demand Levels, Yerevan, Armenia

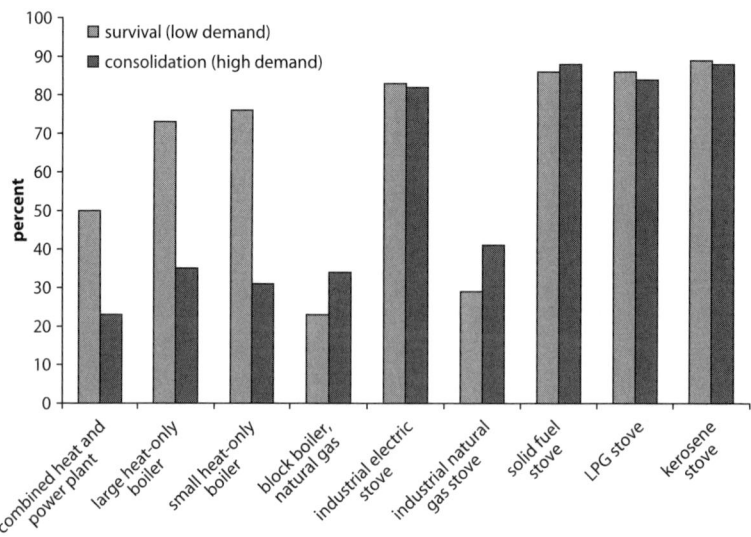

Source: COWI (2002a).

Figure 8.7. Average Cost of Heating for High and Low Demand, Yerevan, Armenia

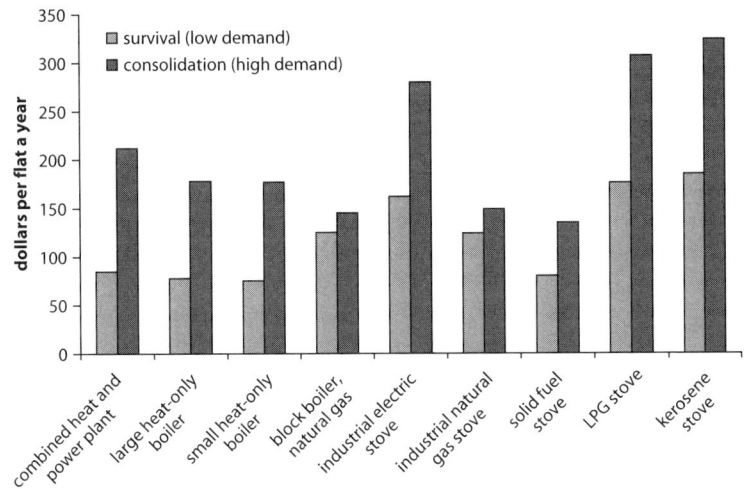

Source: COWI (2002a).
Note: Percentage of population purchasing heating services at different prices: 80 percent at $50 a year, 60 percent at $70 a year, 40 percent at $100 a year.

air pollution and deforestation. In cities where incomes are growing, investments in high efficiency and environmentally benign centralized heating may be justified. If governments choose to invest in centralized options for heating, though, consumers must be able to choose from a range of heating levels with corresponding payment levels so that they are as flexible as individual heating options.

If incomes and heat demand are expected to remain low for the foreseeable future, even a modern, flexible system with lower costs will be unaffordable for many families. In many of the small towns of ECA, district heating systems are in dire need of renovation, with investment requirements beyond the means of these towns. And the high fixed costs of centralized heating systems make them relatively slow to react to a heterogeneous heat demand. In these cases, the best strategy for investments in heating technology are individual systems at the building or apartment level, which may be least cost. If individual options are chosen, more investments are needed in clean and efficient technology to lessen the social costs of traditional fuels.[13]

Other Policies

Whether individual or centralized heating options are chosen, metering and control options are vital so users can choose levels of heat and spending. All centrally provided heat supply options can be fitted with meters and control options that make the systems more flexible. Whether and how much consumers can actually save depends on the level of over- or underheating and the relationship between the system's fixed and variable costs. In general, individual metering and control can save 15–20 percent of heat energy. In some countries where individual meters are not yet in place, a crude approximation of a flexible district heating system has been used. Consumers are allowed to disconnect some of their radiators, and payment is based on the number of radiators in use.

Better insulation of buildings is also necessary to lower the amount of heating required to achieve a minimum comfort level. Most buildings in the region use two to three times as much heat as buildings in comparable climates in Western Europe. However, beyond such basic solutions as fixing broken windows, repair measures can be expensive, typically taking 5–10 years to pay back investments with lower bills.

Conclusion

With socialist-era heating systems in need of repair, new investments are being considered for heating projects across the region. But transition, reform of the power sector, rising prices, and falling incomes have produced the greatest change in demand for heat since the Soviet era. Before making the considerable investments required to rehabilitate district heating systems, it is important to measure the demand for heat against the supply options offered by this and other systems.

On average, the poor spend almost twice as much of their household budgets on heating as do the nonpoor, and they are less income and price elastic than the nonpoor. It should be possible to design a heating subsidy that will benefit the poor more than the nonpoor. But here is the problem often faced with tariff-based subsidies: access. While access to electricity is almost universal in ECA, access to network heating is greater among the nonpoor. Because they have greater access to clean energy networks, the nonpoor will capture the bulk of any subsidy passed through the network, unless the access rate of the poor increases.

But the analysis suggests that extending access to centralized network heating systems may not be appropriate for poor households. The data from Armenia, the Kyrgyz Republic, and Moldova suggest that, unless there are significant improvements in heat quality, poor people are unlikely to pay for heating systems costing more than US$25–$40 a year because they can find less expensive ways to heat themselves. This implies that district heating may be a redundant option for many places and other measures are needed to assist the poor with heating.

The study also highlights the importance of focusing on heating substitutes, generally traditional fuels for the poor. Social costs associated with their use may warrant public intervention, either through increasing incomes or reducing the relative cost of clean fuels through subsidies or investments in efficiency.

The findings in this chapter provide important insights into designing pro-poor heat investments and policies to promote clean choices for the urban poor, depending on local conditions. With its focus on Armenia, this study fed directly into the design of a World Bank urban heating project there; the finding that even flexible and efficient district heat is unaffordable for the majority of Armenia's poor effectively ruled out a large-scale investment in district heating system rehabilitation.[14]

The focus is now on decentralized options. In situations with very different conditions, for example, countries without access to gas, with CHP plants that are indispensable for the power sector, or where it is significantly colder, the policy prescription may be different. And rethinking heat supply must be accompanied by policies to help consumers control their heat consumption and spending (chapter 8). But the implications of this study are wide reaching and highlight the importance of understanding household demand when designing any heating intervention.

Notes

This chapter is based on Lampietti and Meyer 2002.

1. Of 140 infrastructure projects under preparation or implementation in ECA, 20 are either for heat rehabilitation or have a heat component. China has seen a small number of heating projects.

2. Particularly Poland and the Baltics. In the Estonia District Heating Project considerable energy efficiency improvements were achieved: "The Project has made efficiency gains in the areas of heat production, transmission, distribution, and consumption. In the production process, the specific fuel consumption has been reduced by an estimated 5–10 percent, on average. The renovation of the transmission and distribution networks and installation of variable speed pumps has led to significant energy savings, again estimated in the order of up to 10 percent heat and pumping losses. Very dramatic reductions in water losses have also been achieved through the switch from direct to indirect domestic hot water connections, amounting to a decrease of over 85 percent in Tallinn, of almost 90 percent in Tartu, and over 90 percent in Parnu. The heat consumption in buildings equipped with renovated

substations has been estimated to have been reduced by about 24 percent, on average," (World Bank 2000d, p. 7).

3. Direct cash transfers are discussed in chapter 9.

4. This is in contrast to the energy demand model presented in chapter 3. For energy, income elasticity is higher for the poor than for the nonpoor. For heat, the income elasticity of the poor is lower than for the nonpoor.

5. The consumption and expenditure results here are not identical to those in the previous section on household energy demand because the analysis in this section focuses only on a subsample of urban households for which heating information is available.

6. While heat is a public good at the household level, larger (poor) households tend to consume more energy than smaller (nonpoor) households. There are on average two more people in poor than in nonpoor households. Also, there is not much differentiation in living area because commercial real estate markets are not well developed in the sample countries.

7. In Moldova, the difference is not statistically significant at the 5 percent level.

8. Comparative studies ("heat plans") have been carried out in many cities in Eastern and Western Europe confirming this result for greenfield development as well as for modernization of existing district heating systems.

9. The effective indoor temperature would be 20°C, considering 2°C additional from appliances and body temperature.

10. For all heating options represented in figure 8.5, investments have been included to ensure that the equipment would be functional over a lifetime of 20 years. As a result, the costs per apartment are lowest for wood stoves, building-based natural gas boilers, and apartment-based natural gas heaters. But the current natural gas tariff for small consumers is only about 17 percent higher than that for large customers, and so does not reflect the higher distribution costs. The analysis is based on a cash-flow methodology, where all future cash flows are discounted by a discount factor of 10 percent a year.

11. In many countries of the region, however, the already overburdened electrical distribution network would have to be strengthened to cope with additional heat loads. This strengthening would cause additional investments, reflected in higher electricity tariffs.

12. Parts of the centrally supplied district heating system that are not economic to supply must be shut down. Minimum investment plans to make heat supply and consumption more efficient must be devised. Financing sources must be identified. And management and institutional measures to make the remaining least-cost district heating systems viable both for producers and consumers must be identified, which requires rebalancing tariffs between electricity and heat and commercializing the utilities. For details, see Swedpower/FVB (2001) and COWI A/S (2002b).

13. In Georgia and Mongolia, improved stoves for wood and coal have been developed and commercially distributed. These stoves use much less fuel, burn much cleaner, and do not cost much more than a regular, inefficient stove. For Mongolia, see ESMAP (2001).

14. World Bank (2005b), p. 6.

PART 3

Lessons

CHAPTER 9

Implications for Operational Design

The social and political effects of improving utility cost recovery can produce considerable skepticism from stakeholders—as seen with electricity
sector reforms in Armenia, Azerbaijan, Georgia, and Moldova. A sensitive
and well-considered approach to designing policy can thus make a crucial
difference to the sustainability of utility reforms. This chapter looks at the
probable effects of electricity reform in 17 countries in the region that are
at different stages of reform. By understanding how household behavior
will change in response to tariff increases, informed judgments can be
made on what strategies and policies are most likely to be effective in mitigating the welfare losses from reform and how to encourage the poor to
make clean fuel choices.

Simulating the Impact of Tariff Reforms

Household data can be used to simulate the potential effect of raising tariffs to cost-recovery levels.[1] As with the Azerbaijan chapter, such simula-
tions require information about the price elasticity of demand to estimate
consumption following a tariff increase. But empirical estimates of price

elasticity of demand are not readily available and those that are cannot simply be used without careful thought about substitutes, current electricity consumption levels, and the duration of tariff reform.

Figure 9.1 presents a typology of elasticities based on experience and the available literature (annex 5). The key to using the typology effectively is careful thought about local conditions. For example, in a country where gas (or other appropriate substitute) is readily available and inexpensive, where people consume substantially more electricity than basic minimum needs, and where tariffs will be increased slowly, demand is likely to be elastic and the increase in expenditure on electricity in response to tariff increases will be lower. And in a country where substitutes such as gas are expensive, people consume close to minimum needs, and tariffs are increased quickly, demand is likely to be inelastic—and the increase in expenditure will be higher.

The correct measure of consumer welfare loss from a tariff increase is the change in consumer surplus, the gap between the price a consumer actually pays for a good (electricity) and the maximum price he or she would be willing to pay rather than go without it. The larger the consumer surplus, the better off is the consumer. If the price paid increases, for example as a result of a tariff increase, consumer surplus will decrease. The larger the change in consumer surplus as a result of tariff increases, the more acutely the consumer's welfare will be reduced as a result of the new price.

Figure 9.1. Price Elasticity of Residential Power Demand Depends on Local Conditions

		Price of substitutes	
		High	Low
Reform time horizon	Short (1–2 years)	Less than or equal to 0.25	0.75–1.00
	Long (more than 2 years)	0.50–0.75	Greater than or equal to 1.00

Source: Authors' estimates.

Simulating the effect of tariff reform to cost-recovery levels reveals that the change in consumer surplus varies depending on current share of income spent on electricity, the difference between the current tariffs and cost-recovery tariff, and the elasticity of demand (table 9.1). The greatest changes in consumer surplus—the worst-affected consumers—can be seen in Armenia and Serbia. In all countries, the change in consumer surplus is greatest for the poor. Because the poor spend a larger share of their income on electricity, raising tariffs leads to a greater proportionate welfare loss for this group.

The simulation can also be used to identify the cash compensation that would be needed each year to offset the impact of a tariff change (table 9.2). These figures suggest that without effective mitigation measures, the impact of tariff increases may be enough to increase the number of households living below the US$2.15 poverty line.[2]

Of course, as with the Azerbaijan study, calculating the impact of tariff increases without taking into account the effects of rising incomes produces a worst-case scenario. To calculate the income effect, information is needed on how quickly incomes grow and, more importantly, how income growth is distributed. Assuming there will be income growth, small, gradual tariff increases rather than abrupt, large ones will soften the blow to household welfare.

Table 9.1. Percentage Point Change in Consumer Surplus Following Electricity Tariff Increase to Full Cost Recovery

Price elasticity	$e = -0.25$			$e = -0.50$			$e = -1$		
Country	Lowest 20%	Highest 20%	Total	Lowest 20%	Highest 20%	Total	Lowest 20%	Highest 20%	Total
Albania	5	3	4	4	3	4	3	2	3
Armenia	6	4	5	6	3	4	4	3	3
Azerbaijan	4	3	3	2	1	1	−2	−2	−2
Belarus	2	1	1	2	1	1	1	0	1
Bulgaria	3	3	3	3	3	3	3	2	3
Georgia	5	2	2	4	1	2	3	1	2
Kazakhstan	4	2	3	3	2	2	1	0	1
Moldova	3	2	2	3	2	2	2	1	2
Romania	1	1	1	1	1	1	1	1	1
Russia	4	2	3	2	1	1	−3	−1	−2
Serbia	10	6	8	8	5	6	3	2	3
Ukraine	4	2	3	3	2	2	1	0	0

Source: Authors' estimates from household budget survey (HBS).

Table 9.2. Per Household Annual Cash Compensation to Offset Electricity Tariff Change for a Range of Demand Elasticities
(dollars)

Country	e = −0.15	e = −0.25	e = −0.35	e = −0.50	e = −1
Albania	108	103	98	90	65
Armenia	47	45	43	41	31
Azerbaijan	58	48	38	22	n.a.
Belarus	20	19	17	15	8
Bulgaria	71	70	69	67	61
Georgia	23	23	22	21	18
Kazakhstan	43	39	35	29	9
Moldova	12	12	12	11	10
Romania	11	11	11	11	11
Russia	59	48	36	20	n.a.
Serbia	207	190	173	148	64
Ukraine	36	32	29	23	4

Source: Authors' calculations.
n.a. = not applicable

The most important conclusion gained is that households with very low electricity consumption will suffer higher welfare losses from tariff increases because their demand is very inelastic. At the beginning of transition there was scope for efficiency gains from lower consumption because households in the region were traditionally energy intensive due to low residential energy prices. But the move toward cost-recovery tariffs left little scope for further reductions in consumption, and if the price of electricity increases further, high welfare losses will result.

Softening the Blow: Direct Transfers and Lifeline Tariffs

Quantifying the welfare impact of tariff increases does not imply that households should receive full monetary compensation for their welfare losses. This is a choice that needs to be made by the country government, taking into account a multitude of factors that weigh in on this decision. Quantifying welfare impacts is a tool to illustrate to governments the possible tradeoffs between efficiency and equity. Electricity reform is accompanied in most cases by government measures to mitigate the welfare effects of price increases through assistance to vulnerable households. This can be through direct transfers to help with electricity payments, or as a tariff-based subsidy, for example a lifeline tariff where an initial block of electricity consumption, usually up to the minimum basic need, is subsidized by charging it at a much lower rate than subsequent consumption.

The debate on the validity of direct income transfers versus tariff-based subsidies is one of the most contentious in utilities reform. But lessons about the region point to key considerations that can inform good policy decisions.

Ideally, any measures designed to cushion the blow from tariff increases should be well targeted to minimize costs for the government and not lead to price distortions that encourage inefficient resource use. Critics of tariff-based subsidies argue that they are expensive and socially regressive. Since they subsidize the first block of consumption for all consumers, they benefit the poor and the nonpoor, and they encourage inefficient energy use. Opponents of direct income transfers claim that payments through the general social assistance system, while theoretically attractive, fail to reach a large share of the poor because of inadequate targeting. In scoring subsidy schemes against select criteria (coverage of the poor, targeting [the share of the subsidy that goes to the poor], predictability of the benefit, price distorting and other side-effects, and the cost and difficulty of administration), Lovei and others (2000) found that instruments performing well on some criteria performed poorly on others. Not all subsidy mechanisms are applicable or perform equally well across all countries and utility services, and no single instrument has been identified that would outperform all others.

Income transfers tend to be well targeted in countries with a small percentage of the population below the poverty line. In this case, as long as there are enough funds to finance the administration of social assistance and the informal sector is small, means testing is easy; examples include Hungary and Poland. It is harder to produce well-targeted income transfers in countries where nearly half the population is poor, budget resources are insufficient, and means or proxy means testing is very difficult because of a large informal sector.

A key problem in Europe and Central Asia (ECA) is that social protection systems and energy-specific safety nets are not well correlated with poverty. In the past they were based on categorical privileges of the kind seen in the Moldova chapter. And reformulation of categories can be politically difficult, time consuming, and expensive. The amount of compensation is often subject to political exploitation. Improvements in targeting are being made, but this takes place over several years. In the meantime, direct transfers can be as wasteful as tariff-based subsidies, as seen in the Georgia study.

Furthermore, coverage of the poor is inversely related to the share of the subsidy that goes to the poor; the more households targeted by the assistance, the more likely households that do not fulfill the poverty criteria are assisted (table 9.3).[3] If a benefit system covers a large percentage of the population, it is likely that it is poorly targeted.

The case for lifeline tariffs is stronger in countries with high poverty rates, high inequality, high access of poor households to the subsidized network, and poor targeting of social transfers. The greater the number of poor people, and the higher the rate of the poor who have access to the subsidized network, the higher the coverage and the lower the leakage of lifeline tariffs. But there should be sufficient political will to keep the lifeline tariff blocks small (below 50 KWh or 100 KWh), and the government must compensate utilities for any social transfers they provide.

Going forward, the choice of instruments must be determined on a country-by-country basis in careful consultation with the client country, with consideration given to the percentage of the population below the poverty line, the available budget, and the timeline for reform. Policy makers in countries with high poverty rates may find lifelines a more efficient way to deliver mitigating measures than direct income transfers channeled through questionable social protection systems. In cases where strong vested interests are opposed to tariff-based subsidies, the use of pilots to introduce change can be effective.

Comparing the ratio of benefits to costs for each program provides a measure of the efficiency of the transfer. As noted earlier, the change in consumer surplus approximates to the amount of money that would need to be given to a household to offset the impact of a tariff change. A larger change in consumer surplus points to a greater negative effect from a tariff increase. Multiplying this number by the number of poor (below the US$2.15 poverty line) approximates the budget for a cost-effective

Table 9.3. Leakage and Coverage Are Highly Correlated, 2002
(percent of population)

Country	Coverage of the poor (below the US$2.15 poverty line)	Share of subsidy that goes to the poor
Armenia	28	65
Azerbaijan	61	7
Bulgaria	27	33
Kyrgyz Republic	21	92
Poland	87	6
Ukraine	11	7

Source: Authors' calculations based on household survey data.
Note: Leakage is the proportion of people reached by a given program who are nonpoor. Coverage is the proportion of the poor in a society who are reached by a program.

mitigating program. Dividing this figure by the number of people bene-fiting from the lifeline and by the number receiving the direct income transfer gives the average benefit per person (poor or not) for each program. This figure will always be higher for income transfers, because the number of people receiving electricity is higher than the number that receive social protection assistance. For lifelines, as the incidence of poverty increases so does the efficiency of the lifeline.

Another key consideration is timing. While a poverty-targeted income transfer is more efficient, it may take years to become operational. Thus, in the near term, the only feasible solution may be to channel compensa-tion through the existing social protection system. Ideally, tariff-based sub-sidies should not be phased out until targeting is significantly improved.

Other Considerations

It is possible for lifelines to be self-funding. But this requires setting the tariff for the lower blocks below cost recovery and the higher blocks above it, possibly resulting in inefficient resource consumption. It also places the burden of financing the subsidy on the utility and consumers with higher consumption, rather than the government.

Alternatives to direct transfers and lifeline tariffs can be explored.[4] A common objection to lifeline tariffs is that they are socially regressive and wasteful because they subsidize the first block of consumption of all consumers, poor and nonpoor. A volume-differentiated tariff avoids this problem. It works by charging a lower tariff for households consuming less electricity than households consuming above a certain threshold level. Households consuming above this threshold level—which, according to HBS data, tend to be those on higher incomes—are charged a higher rate for all their consumption. Thus nonpoor households are unlikely to receive any of the subsidy.[5] As with lifeline tariffs, appropriate measures should be taken to avoid incentives to game the system.

Any kind of tariff-based subsidy cannot work in the absence of meters. Many, but not all, countries in the region have good residential metering for electricity; lack of meters is more problematic in district heating, gas, and water. It is true that if lifelines are to be introduced in these other sectors—and there is a strong rationale for this—the cost and time required to introduce meters becomes a major challenge. But meters are important for other reasons too. Without meters it is impossible to measure or estimate the potential impact of tariffs on consumption and payment patterns; it is impossible to see whether direct transfer payments are targeted at low consumption households and are at the right level; and where there are no meters, and billing is based on average consumption, incentives to conserve are weak.

Where tariff-based subsidies are in use, it may be possible to reorient their design to maximize consumer welfare gains and minimize the cost to the government budget, as was done with the simulation of an alternative subsidy design aimed at households consuming within a certain margin in the Georgia study. Another consideration is that using a price-based instrument can carry a positive externality if it encourages use of clean fuels. As seen in chapter 8, if traditional fuels are significantly cheaper than clean fuels, poor consumers may choose to spend direct transfers on consumption of traditional fuels that carry social and environmental costs. The poor must have access to network energy for the benefits of lifeline tariffs to reach them; if not, the bulk of energy subsidies will go to the nonpoor.

Other Pro-Poor Mitigating Measures

In addition to income transfers and lifelines, a number of other actions can be taken to shield the poor from higher tariffs.

Explicitly Link Tariff Increases to Improvements in Service Quality

As noted earlier, there could be a mismatch between the timing of the costs (higher tariffs) and benefits (improved service quality) of tariff reform. In this case, the welfare loss from raising tariffs can be minimized by explicitly linking tariff increases to improved service quality, particularly important for poor people who often suffer from the lowest quality service. It is also likely to generate more political will to support the reform.

But cost recovery is a requirement for the investments that will improve service quality, so in most cases there is a time lag between higher tariffs and tangible improvements in service quality. The exceptions will be countries that can afford to time tariff increases on the basis of political considerations, such as Azerbaijan. In Georgia, this option was theoretically possible and the utility attempted to use full services as an incentive to pay bills, but the utility was unsuccessful due to political interference in electricity dispatch.

It is often difficult to quantify service quality improvements. Limited aggregate data suggest that service quality improved in a number of capital cities. But identifying any of the benefits of reform in the region is confounded by changes in record keeping and accounting methods, by vested interests, and by private sector operators with few incentives to report production efficiency gains. Data on service quality, even of a very basic nature, are seldom available. Measuring the number and location of blackouts over time, for example, and how dependent households in blackout areas are on electricity (less in rural areas than in small towns), would provide a fairly straightforward way of measuring the impact of return to 24-hour service. Other indicators that can be monitored to ensure better service quality include number of outages and frequency and voltage stability.

Looking forward, World Bank operations can improve transparency and accountability by emphasizing a systematic set of indicators in all sector operations and by disseminating this information to the public. A best practice example is the Armenian Natural Monopoly Regulatory Commission, which discloses monthly power sector performance indicators on the Internet.[6] A system of citizen feedback on service delivery, similar to the public services report cards used in the Philippines and India,[7] can be instituted.[8] Such a mechanism can create a direct link between service quality and tariff increases. Another good illustration is provided in the World Bank's *World Development Report 2004: Making Services Work for Poor People*, which focuses on how to make basic services—health, education, water, sanitation, and electricity—more accessible for the poor. The report outlines a system of accountability that connects consumers, government, and providers through four interrelationships: improving "client power" by making utility providers accountable to the poor, increasing the voice of the poor, improving compacts between policy makers and service providers, and instituting better management procedures.

Raise Tariffs Slowly

The shock therapy programs of the 1990s included sudden, radical increases in tariffs. Sudden changes in tariffs require people to change their behavior very quickly, which is not always possible. Raising tariffs slowly minimizes welfare losses by allowing consumers to adjust their consumption patterns, take advantage of income growth, and increase use of substitutes. But it is also likely to have significant fiscal costs, especially if tariffs are well below cost recovery. So, if sudden, large tariff increases are absolutely necessary, they should be accompanied by programs that provide households with the resources necessary to adjust to the new tariff structure.

Raise Collections First

Improving cost recovery requires that tariffs be set to appropriate levels and that these tariffs be enforced. But because nonpayment tends to be higher among the poor before reform, increasing enforcement of tariffs alongside price increases will lead to larger effective tariff increases for the poor than the nonpoor, as seen in the country studies. To accurately calculate the impact of tariff increases, policy makers should consider the price effect of increased tariff enforcement, which can create a much larger de facto increase than predicted. Unless efforts are first made to raise collections, the poor will cope with tariff increases by nonpayment or disconnection.

Increase Access to Gas or Other Clean Substitutes

As seen in chapter 3, the poor generally have less access to gas infrastructure, making it harder for them to reduce electricity consumption by substitution of more efficient alternatives. At the time, increasing access to such clean and inexpensive substitutes as gas might have been one of the best ways to offset the impact of electricity tariff increases, particularly where a large number of people heat with electricity (figure 9.2). But the true cost of gas is often higher than reflected in the figure. Many countries, such as Armenia and Romania, keep gas tariffs below the true economic cost, distorting consumer choices away from district heating to gas-fired heating.

The recommended switch to gas must therefore be qualified in light of recent signs that Russia's willingness to supply large volumes of subsidized gas may be coming to an end. If this happens, the price of gas will increase

Figure 9.2. Electricity Tariffs are Higher Than Gas Tariffs, 1992–2002

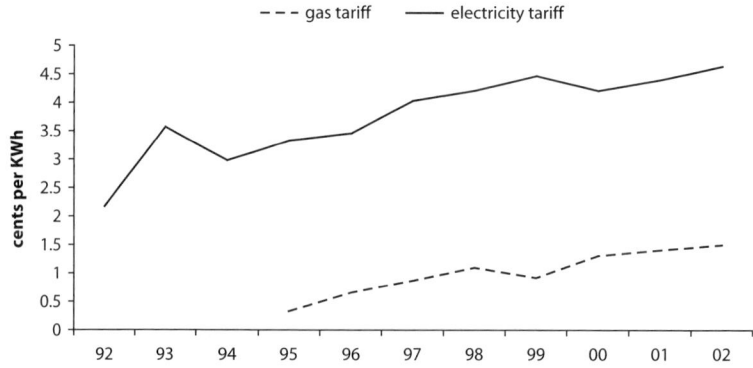

Source: Authors' calculations based on data from local consultants, Counterpart International (for Moldova), and ERRANET database.
Note: The applied conversion factor was 277.8 KWh per Giga-Joule of natural gas (International Energy Agency). Average tariffs were calculated for Armenia, Azerbaijan, Georgia, Hungary, Kazakhstan, Moldova, and Poland. This figure is a simple average. The number of observations varies by year depending on data availability.

significantly from the levels indicated in figure 9.2. Subsequent changes in relative prices must be taken into account when looking at the costs of electricity and gas. One way to increase access to gas, if this policy is chosen, is for the government to bid out competitive subsidies to encourage extension of natural gas networks to poor neighborhoods.

Make Metering a Priority

In an environment of tariff reform, meters offer consumers information about and control over their energy use, leading to savings and possibly to more efficient consumption—for electricity and for other network energy sources, including gas and district heat. Whether and how much consumers actually save depends on the level of over- or underconsumption and the relationship between the system's fixed and variable costs.[9]

Imaginative use of technology can make meters a more helpful instrument for consumers to control their expenditures, for example, through "smart" metering technology. The simplest form of smart metering is a display meter that allows consumers to monitor consumption in money terms rather than kilowatt hours (KWh). It can be combined with a keypad or smart card reader linked to prepayment systems, potentially reducing costs and allowing consumers to take advantage of lower tariffs generally offered for prepayment. Internet-linked systems can offer other services, including direct welfare benefits payments. Realizing the full potential for smart metering requires piloting the technology to establish the real value to customers. On the downside, it is unrealistic to expect low-income households to meet the cost of installing expensive new systems.[10]

Investments in Efficiency

Most buildings in the region use two to three times as much heat as buildings in comparable climates in Western Europe. Particularly promising for reducing energy expenditures, especially in areas where large increases in clean fuel prices are expected, are investments in efficiency and insulation that can produce substantial reductions in consumption.

Financing Instruments

Investments in efficiency and access to gas often carry high initial costs and must be coupled with innovative financial instruments that enable consumers, particularly the poor, to distribute capital costs over a longer period. As seen in Georgia, focus group participants said that the costs of connecting and appliances were barriers to installing gas. Financing instruments can help defray recurring costs of higher energy prices, as illustrated in the Armenia study, where customers commonly paid off their higher winter electricity bills during the summer months.

Mitigating the Environmental Effects of Reform

Increased production efficiency, new investment, and environmentally friendly technology accompanying reform were expected to contribute to lower fossil fuel consumption and lower emissions, better ambient air quality, and thus to better health outcomes for the local population.[11] But reforms affected household fuel choices, which also carried environmental effects. This section looks at the environmental impact of sector reform and its impact on poverty.

Environmental Benefits from Increased Energy Production Efficiency?

Claims about improvements in ambient air quality because of reforms are difficult to verify for most pollutants, since pollutant indicators and monitoring programs were never established in ECA—or if they were, collection collapsed with the breakup of the Soviet Union and the subsequent transition. Measured by fuel efficiency of electricity production, the environmental performance of the electricity sector has improved slightly over the past decade, leading to reductions in carbon dioxide emissions and positive impacts on global and long-range air pollution.

Evaluating these benefits requires sophisticated climate change models well beyond the scope of this book; in any event, the benefits are global rather than local.[12] In most cases, increasing energy efficiency in electricity production has little direct impact on human health, because the electricity sector's share of total health damage from air pollution is negligible. Moreover, it does not contribute greatly to the pollutants that cause the most local health damage.[13] If power plant stacks are high or located in sparsely populated areas, as in much of the region, they may not have much influence on ambient air quality.[14]

If the sector does not help determine local air quality, reforms will produce small health benefits even if emission reductions are large. The raw data suggest that urban air pollution decreased slightly in the major cities during the reforms, though it continues to be a health hazard. [15] How much did the power sector reforms contribute to this change? A crude dispersion model was used to estimate the magnitude of the impact of the sector on air quality and health in selected cities.[16] The model found that the power sector contributed less than 1 percent of health damage from all emissions because of its low contribution to total emissions. Between 1990 and 2000, the share of the electricity sector in the disability adjusted life years (DALYs) originating from low air quality ranged 0.1–2.0 percent.[17]

The analysis reveals five reasons for the sector's low contribution to health damages:

1. The substantial drop in the amount of electricity produced in Armenia, Georgia, and Kazakhstan.
2. The shift in the fuel mix used for thermal power plants toward natural gas in Armenia and Azerbaijan.
3. The location of high-capacity power plants far from populated cities.
4. Improvements in fuel quality[18] and abatement technologies for particulate matter that were already in place before the reforms started in Hungary and Poland, with average removal efficiencies of 97–99.9 percent.
5. The fact that power station stacks were built high to reduce deterioration of ambient air quality and were regulated by Soviet norms and regulations.

The share of overall emissions from power stations is falling as private transport has become a major source of urban air pollution in the large cities of ECA.[19]

Environmental Costs from Fuel Substitution

While the environmental benefits of increased production efficiency are fairly ambiguous, the analysis confirms the findings of the country case studies: there may be unintended environmental costs associated with reforms. As residential tariffs are brought to cost-recovery levels, households, particularly in low-income groups, may switch to cheaper traditional fuel (wood, coal, or kerosene), which contributes to indoor and outdoor air pollution. Although there are no comprehensive data on household emissions, survey evidence on household substitution behavior does exist. In the Armenia study, for example, 80 percent of households and 95 percent of poor households reported using alternative fuel sources (primarily wood) to reduce reliance on electricity. And a report by the United Nations Environment Programme (UNEP) indicated rising air pollution because of increased low temperature emissions, a large share of which is attributable to household heating.[20] In Katowice, one of Central

Europe's most severely polluted cities, the primary source of local air pollution is household burning of coal for heating.[21]

Health damage from burning traditional fuels may be substantial and
may exceed the benefits from reduced power plant emissions, especially
in densely populated urban areas where household chimneys are low and
there is little opportunity for pollution to disperse. The dispersion model
developed earlier, with assumptions about household fuel use, estimates
the share of air pollution attributable to household wood and coal use.[22]
The share of DALYs attributable to households using traditional fuels
ranges between 6 percent and 39 percent over the last decade (figure 9.3),
considerably higher than the contribution of the electricity sector
(0.5–2.4 percent).

Burning traditional fuels can also cause indoor air pollution, which leads
to disease and loss of DALYs.[23] Back-of-the-envelope estimates of the possible maximum extent of health damage from indoor air pollution in three
cities in the Caucasus put the number of premature deaths at the same
order of magnitude as that from outdoor air pollution (table 9.4). But
more research is necessary to identify the relationships between fuel use
(including technology and chimney availability) and indoor air pollution
and health outcomes. The number of premature deaths is higher among
women than children under age five.[24] The total estimated potential loss of
life because of indoor air pollution amounts to 7 percent of all deaths
related to respiratory diseases and 1 percent of the total deaths in Armenia,

Figure 9.3. Electricity Is a Small Share in Health Damage
(average 1990–2000)

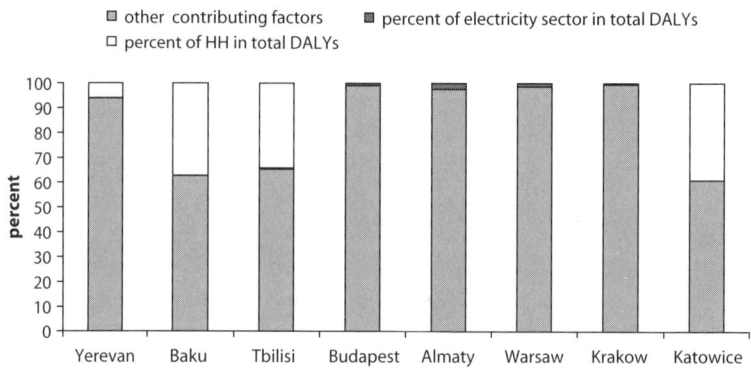

Table 9.4. Potential Maximum Loss of Life and Life Years from Indoor Air Pollution

	Armenia (Yerevan)	Georgia (Tbilisi)	Azerbaijan (Baku)
Number of premature deaths			
Children under age five	52 (19)	62 (20)	36 (17)
Women	164 (60)	147 (47)	114 (54)
Total	**216 (79)**	**210 (66)**	**150 (71)**
DALYs			
Children under age five	1,820 (664)	2,186 (690)	1,260 (597)
Women	3,287 (1,199)	2,928 (931)	2,275 (1,078)
Total	**5,107 (1,863)**	**5,134 (1,621)**	**3,535 (1,675)**

Source: Authors' calculations based on World Health Organization statistics, mortality database, and household surveys.

10 percent and 1 percent in Georgia, and 2 percent and 0.3 percent in Azerbaijan.[25]

Electricity Reform and Deforestation

Fuel wood use may also contribute to deforestation and the loss of important forest resources—though the difficulty of obtaining data on deforestation, particularly that attributable to fuel wood collection, makes it unclear whether this is a problem. Several studies and observations by forestry specialists visiting the Caucasus show a significant decrease of local forest cover and deterioration in forest quality, but these trends are often not reflected in national or international statistics.[26] Indeed, the forested area appears to be increasing from 0.2 percent a year in Poland to 2.2 percent a year in Kazakhstan in the last 10–15 years,[27] although these data are unlikely to be reliable since few ministries have the resources to monitor forest cover consistently and rigorously.[28] It may also be that there is no visible change in total forest cover but the quality and density is decreasing.[29]

Certainly the low intensity harvesting of fuel wood from trees growing in agricultural land, around houses, and along roads is seldom shown to have a significant impact on overall forest canopy cover (and is difficult to measure with remote sensing). And when trees are coppiced or pollarded to provide these supplies, the overall impact of rural firewood harvesting can be negligible. But the situation is quite different in meeting urban demands for firewood. When urban household energy use is constrained because of utility reform, the negative environmental impact on forested areas can be significant because of a shift from electricity to firewood, which can create market conditions that favor clear cutting of large forested areas. From the household surveys, it is clear that the majority of rural households and a substantial number of urban households used wood for their energy needs. An average expected household consumption of 5–10 cubic meters of fuel wood per year can lead to substantial local deforestation. More research is necessary on the amounts of fuel wood that households burn, the sustainability of this practice, and the incremental use of fuel wood resulting from electricity reforms. In addition it must be remembered that poor electricity supply prior to reform also creates conditions favoring deforestation, as with Armenia before reform.

How to Improve the Environmental Effects of Reform

Better monitoring of ambient environmental quality improvements is necessary for future measuring of the environmental effects of reform. Better information is also needed on fuel substitution to evaluate the impact of reforms on fuel switching, energy use, substitution effects, and health and social effects. Household surveys currently do not reveal enough information about energy and other utility reforms and need to include questions about utilities.[30] Developing models to help predict behavior under a variety of scenarios is also necessary.

It is likely that reforms have damaged health because households switched to traditional fuels. One solution is to improve access and efficiency in using clean alternatives. Survey data indicate that fewer households would use wood and coal if they had access to gas. Of course, in many countries the gas sector is also in need of reform before it can operate on a sustainable basis. In cases where switching to traditional fuels is a problem and where the poor overwhelmingly have access to gas, another option is to use a tariff-based subsidy rather than a direct transfer to encourage the poor to make cleaner fuel choices.

Notes

1. In this chapter, the same household data are used as for the energy demand analysis in chapter 3, which are from 17 of 29 ECA countries. Detailed Household Budget Survey analysis can be found in Annex 2. As noted earlier, consumers gain from an improvement in service quality and the removal of rationing, but lose from tariff increases and disconnections. Existing data collection efforts have not focused on quality, so at this time the gains from the service quality improvement cannot be measured. Therefore, the focus of the remaining analysis is on the consumer surplus change from a price increase.

2. While difficult to determine in practice, the correct measure of compensation would be the amount of cash the consumer would need, given the new tariff structure, to be as well off as before the tariff change.

3. Leakage is the proportion of people reached by a given program who are non-poor. Coverage is the proportion of the poor in a society who are reached by a program.

4. For further reading on direct transfers and lifeline tariffs, see Komives and others (2005), chapter 6.

5. For more information on volume-differentiated tariffs, see Komives and others (2005), p.13. Alternative pricing methods and their distributive implications are also discussed in Linn and Bahl (1992). Multidimensional tariff schemes can differentiate between capital and variable costs and increase the scope for designing targeted subsidies.

6. Web address: http://rcnm.am. View indicators under "sector reports" link.

7. Bhatnagar (2001); Paul (1994, 1998).

8. Municipal utility users' feedback was a key feature of the World Bank-Supported People's Voice project in Ukraine (http://web.worldbank.org).

9. For district heat, individual metering and control can save 15–20 percent of heat energy.

10. There is some tension between the advantages of effective individual metering and the experience of private operators and management contractors. In Georgia, communal metering proved an effective way to improve collections at a lower cost than installing individual meters.

11. An analysis of six reforming countries from across the region (Armenia, Azerbaijan, Georgia, Hungary, Kazakhstan, and Poland) was conducted to test the validity of these claims. This material was first published in Lampietti ed. (2004).

12. Since these benefits are global rather than local, the costs could be partially financed through such institutions as the Global Environment Facility or the Prototype Carbon Fund. Costs then would not be borne by the local population, who experience only a small share of the benefits.

13. These are sulfur dioxide, nitrous oxides, and fine particulate matter (PM_{10}). Improvements in air quality resulting from increased efficiency can be insignificant when compared with emissions of these pollutants. Different pollutants are associated with different health risks, commonly measured in terms of the disability adjusted life years (DALY), used internationally to compare health effects of different causes. One DALY is equal to the loss of one healthy life year.

14. In the former Soviet Union, a number of state norms and rules regulated the height and design of power plant chimneys. These rules and norms were

generally close to western norms. The Ekibastuz Power Plant in Kazakhstan, which uses coal as fuel, has two stacks of 330 meters each. Other known stacks in Russia and Ukraine range from 250 meters to 1,370 meters.

15. Ambient air quality standards for PM_{10} continue to be surpassed regularly in Katowice, Tbilisi, and Yerevan (Lampietti and Meyer 2002). Baku, Tbilisi, and Yerevan used to be included in the list of most polluted cities of the former Soviet Union because of the industrialization and urbanization of the past 30 years. Lack of monitoring data precludes in-depth assessment of the state of the air quality, though air quality has been monitored in all the countries for many years. After decentralization, lack of funds and obsolete monitoring methods inhibited progress, and data collection has declined sharply.

16. Details of the model can be found in Lampietti ed. (2004).

17. The highest shares attributed to the sector are in Almaty and Warsaw. In the other cities, the contribution is less than 0.5 percent. Total DALYs from ambient air pollution range from around 4,000 on average in Krakow to around 50,000 on average in Katowice.

18. Sulfur and ash content of coal; sulfur content of liquid fuel.

19. In Tbilisi, for instance, transport accounts for 80 percent of total air pollutants. Private transport emissions are increasing because of the aging vehicle fleet, the low quality and high sulfur content of the fuel, and the decline in public transport.

20. UNEP (2002).

21. Bucknall (1999); Bucknall and Hughes (2002). In the Katowice area, annual average levels of sulfur dioxide exceed the current European Union standard nearly threefold, and annual average levels of PM_{10} are well above the standard. Some parts of the metropolitan area exceed daily PM_{10} limit values for 200 days a year, causing significant respiratory illness and other problems for the population. The largest part of the average exposure to PM_{10} comes from household boilers, responsible for 80 percent of exposure to harmful particles in Katowice voivodship. Power and district heating plants contribute little to exposure because—as a result of their high stacks—their emissions are dispersed over a wider area. The ambient air pollution impact in Katowice is attributable to coal and wood use for heating and cooking purposes, not solely as substitution for electricity. In Katowice, 409 households burn coal for heating, with lower income households more likely to heat with coal.

22. The following assumptions are made: the share of population using wood is as reported earlier, the urban exposed population of Baku is estimated at 50 percent, the average quantity of wood per household is 8 cubic meters per year, the density factor of wood 0.5 ton per cubic meter, and the average household size is taken from UN World Prospects Population Database. In the

2002 census of Georgia, the actual population appears to be smaller than originally listed in international databases of the United Nations and the World Bank. The contribution of the household sector to atmospheric air pollution fell from 34 percent to 29 percent on average for the past 10 years. The relative contribution of the electricity sector to health effects will accordingly be lower, since it also depends on the number of people affected.

23. Worldwide, inhalation of smoke from combustion of solid fuels causes about 36 percent of lower respiratory infections, 22 percent of chronic obstructive pulmonary disease, and 1 percent of trachea, bronchus, and lung cancer (WHO 2002). It is also associated with tuberculosis, cataracts, and asthma (though the evidence here is weaker). Nearly 3 percent of DALYs worldwide are attributed to indoor smoke, 2.5 percent for men and 2.8 percent for women.

24. Though this is counterintuitive given the evidence in other continents, in ECA, the number of women compared with the number of children is higher than elsewhere. In Azerbaijan, it is 6 women per child; in Georgia, 10; and in Armenia, 8.5.

25. These estimates are based on the assumption that ventilation is lacking at the location of traditional fuel burning. More research is needed to identify the availability and use of fuel burning technologies and chimneys in households to establish the precise relation between traditional fuel use, indoor air pollution, and health outcomes.

26. For example, UNEP (2002).

27. Reference periods for the different countries are: Armenia 1983–96, Azerbaijan 1983–88, Hungary 1990–96, Kazakhstan 1988–93, Poland 1987–91 and 1992–96, and Moldova 1990–95. No information is available for Georgia.

28. The United Nations Economic Commission for Europe and Food and Agriculture Organization report (2000) on forest resources of Europe does not indicate a decline in forest resources. A forest is composed primarily of indigenous (native) tree species. Natural forests include closed forests and open forests (at least 10 percent tree cover). Total forests consist of all forest area (plantations and natural forests) for temperate developed countries.

29. Changes in forest canopy cover, which can be monitored using conventional remote sensing approaches, often do not reflect changes in forest health, yield, species mix, or density, which can be captured only by more rigorous ground-based inventories and assessments (UNEP 2002). The Caucasus Environmental Outlook reports that selective cutting occurred when the highest quality trees were cut. During the past 10 years, cutting was extensive on the Saguramo-Yalon range (East Georgia), and on the outskirts of Tbilisi and Yerevan. In state-owned forests, there were no significant changes in the

total cover, but all valuable specimens of beech and some other species have been cut, drastically reducing forest quality. It is estimated by the Caucasus Environmental Outlook that in Armenia and Georgia, 26 percent of beech forest has been converted to coppice forests and only about 10 percent of the beech forests left have a high density.

30. Suggested questions are presented at http://wbln0018.worldbank.org/esmap/site.nsf/pages/Flagship_2006.

CHAPTER 10

Conclusion: Designing Reforms to Produce Better Outcomes for the Poor

The Soviet legacy left Europe and Central Asia's (ECA) power sector in a state of disrepair and dependent on a complicated system of fiscally unsustainable budget transfers. In many countries the result was a collapse of energy utilities and an inability to supply power for normal social and economic activity. To put the power sector back on its feet, governments across the region undertook far-reaching sector reforms, unbundling vertically integrated utilities, liberalizing and regulating the sector, privatizing companies, setting prices at cost-recovery levels, and improving payment discipline. But as with all utility reforms, policy makers were faced with the mismatch between the timing of costs and benefits associated with reform, exacerbated by expectations rooted in communist times that the state would take care of utility provision. Despite attempts to soften the blow, the negative effects of increasing tariffs and collections have often been highly disruptive, threatening the sustainability of reform.

After a decade of reform, most countries have only partially achieved cost recovery, and further tariff reform is needed for much of the region. The analysis and findings of the studies in this book provide information on the expected household responses to reform—and on how the design of reform can be modified to produce better outcomes and mitigate the more negative effects. This chapter provides an overview of the book's key findings on the effects of reform on the poor, effectively mitigating strategies, approaches to designing successful reforms, and methods of analyzing reform.

Tariff Reform: Where Do We Stand?

Estimates indicate that residential electricity tariffs are below cost recovery in 14 of 19 ECA countries (figure 10.1). The largest percentage increases needed are in Central Asia (Azerbaijan, the Kyrgyz Republic, Tajikistan, and Uzbekistan) and in Southeastern Europe (Albania, Macedonia, and Serbia and Montenegro). In absolute terms, Albania, Macedonia and Serbia and Montenegro all need to increase tariffs more than 2 cents per kilowatt hour (KWh).[1] Such sizable tariff increases are unlikely to be welfare neutral unless accompanied by substantial and visible improvements in service quality or cushioned by income transfers.

Figure 10.1. Electricity Tariff Reform Is Still Needed

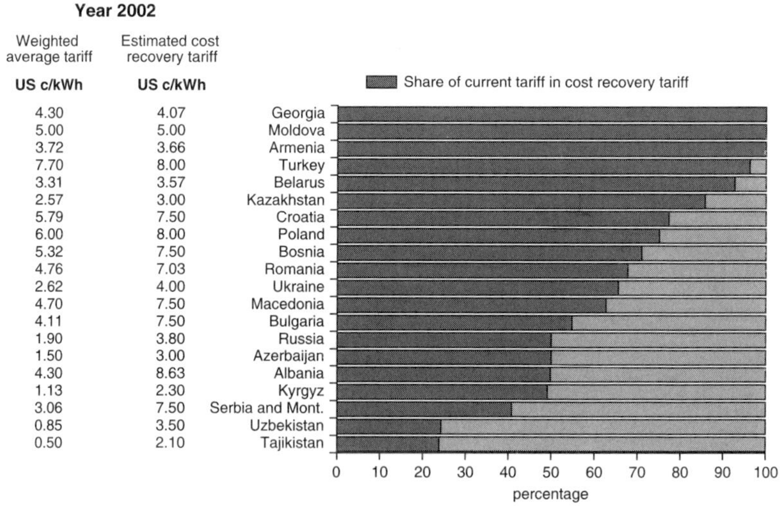

Year 2002

Weighted average tariff	Estimated cost recovery tariff		
US c/kWh	US c/kWh		Share of current tariff in cost recovery tariff
4.30	4.07	Georgia	
5.00	5.00	Moldova	
3.72	3.66	Armenia	
7.70	8.00	Turkey	
3.31	3.57	Belarus	
2.57	3.00	Kazakhstan	
5.79	7.50	Croatia	
6.00	8.00	Poland	
5.32	7.50	Bosnia	
4.76	7.03	Romania	
2.62	4.00	Ukraine	
4.70	7.50	Macedonia	
4.11	7.50	Bulgaria	
1.90	3.80	Russia	
1.50	3.00	Azerbaijan	
4.30	8.63	Albania	
1.13	2.30	Kyrgyz	
3.06	7.50	Serbia and Mont.	
0.85	3.50	Uzbekistan	
0.50	2.10	Tajikistan	

Source: World Bank ECA Electricity Data, 2003.
Note: Residential tariffs are usually set higher than average weighted tariffs because of the higher costs of supplying electricity to low-voltage consumers. So the shortfall in current residential tariffs is slightly higher than represented here.
Russia: no residential tariff available; weighted average end user tariff used instead.
Bosnia and Serbia and Montenegro: Figures are from 2002.

How Do Reforms Affect the Poor?

By using quantitative data to look closely at household behavior, the studies in this book teach us about residential consumption of energy, what different fuels are used for, and what happens when relative fuel prices change. Combined with qualitative data, they help clarify why people make certain choices—for example, whether households use nonnetwork fuels because they do not have access to network fuels or because they cannot afford network fuels.

Ideally, poverty and social impact analyses (PSIAs) can be conducted to analyze reforms before they take place and to model the effects of different policies so that policy makers can make empirically informed choices. But even without conducting new studies, the knowledge and lessons gained from the PSIAs presented here will be useful for future reform in ECA and elsewhere. As noted in the World Bank's guidelines for PSIAs, "where information is sparse and time short, the core issues may have to be addressed on the basis of knowledge of the country and international experience of similar reforms."[2] What lessons can be drawn from these studies for future reform?

Residential Energy Consumption
In a region of low incomes that are only now on a moderate growth path and are not keeping pace with price increases, the poor consume less energy than the nonpoor but spend a higher percentage of their monthly expenditure on energy. Electricity frequently forms the bulk of this energy expenditure.

Nonpayment: Affordability versus Free-Riding
Nonpayment, stemming in part from a legacy of extremely low nominal tariffs, has proved an intractable problem for countries introducing reform. Identifying which groups do not pay enables a disaggregation of affordability and free-riding as possible causes. Though a culture of nonpayment was pervasive in such countries as Georgia, in many cases it is the poor who accumulate the greatest arrears, indicating that the move to cost recovery has resulted in tariffs that are too high for households in lower income quintiles to afford.

Elasticity of Electricity Demand
In response to increasing prices, the poor displayed greater elasticity than the nonpoor—their consumption decreased more rapidly. But with

sustained price increases, consumption levels in some countries are now so low among poor households that they are extremely inelastic. Below a certain level of electricity consumption—typically about 125 KWh per month, enough for a refrigerator and three lightbulbs—it is extremely hard to reduce consumption any further. ECA's cold winters make the need for energy for heat particularly inelastic. Further moves to cost recovery in countries where a significant number of poor households consume at this level need to pay particular attention to measures that will prevent significant welfare losses among the poor.

Coping Mechanisms

Households use various coping mechanisms to reduce their energy expenditures. They include using energy more sparingly (turning off lights) and substituting cheaper fuels (gas and wood) for more expensive ones. While many ECA countries were very energy intensive before reform, allowing some scope for improved efficiency, some of these measures to reduce expenditures can carry significant negative effects that must be factored into calculations of reform outcomes. These include cases where households must go to extreme lengths to conserve energy, such as turning off refrigerators for days at a time, and where such substitutes as wood and coal contribute to indoor and outdoor air pollution, with adverse outcomes for health and the environment. In addition, the studies typically found that the urban poor experienced the most difficulties in coping with increased tariffs, since they face more difficulty than the rural poor in getting wood, a relatively cheap substitute for electricity for heating.

Improvements in Service Quality

For countries with low access rates, a major benefit of reform is better access and service quality. In ECA, the decapitalization of energy utilities and the ballooning energy-related debts limited supply and resulted in electricity rationing. So the possibility of service quality improvements represented a significant potential benefit of reform. For many consumers, the welfare gains from service quality improvements can balance the welfare losses from tariff increases. In countries where service quality improvements did not uniformly accompany tariff increases, as for some consumers in Georgia, not only did welfare decrease, but the support for reform and the willingness of customers to pay were also compromised. This proved a very real threat to the sustainability of reform efforts.

Designing Effective Mitigating Strategies

Direct Transfers or Tariff-Based Subsidies?

How best to mitigate the impact of cost recovery has been the subject of considerable debate between the relative merits of tariff-based subsidies and direct, targeted cash transfers. Lump-sum transfers are usually the most efficient way to help the poor, according to public finance theory and studies looking at parts of the world where the poor have less access to utility infrastructure, and where the nonpoor therefore capture the benefits of subsidies. But when the social protection systems for channeling direct transfers are not well targeted to the poor, as in Georgia, they too can be inefficient, costly, and regressive. In the real world, and with different circumstances, second-best solutions—tariff-based subsidies—sometimes make more sense.

In deciding between different options to assist the poor, governments need to carefully consider the various factors (outlined in chapter 9) that determine whether tariff-based subsidies or direct transfers will be more effective in reaching households in the lowest income quintiles. These factors include the levels of access among the poor, the percentage of poor in the local population, and the targeting effectiveness of existing transfer schemes. In a country where poverty is widespread and where the social benefit system is not well targeted to the poor, a tariff-based subsidy can be a more effective and reliable instrument than a direct transfer through the social benefit system. In Latin America and Africa, tariff-based subsidies have been found to be socially regressive because the poor do not have access to the network and cannot therefore capture the benefits of such subsidies. This is much less the case in ECA, where almost all households, poor and nonpoor, enjoy access to network energy.

Improving the Efficiency of Energy Consumption

Governments can also help mitigate welfare losses by helping households move toward more efficient energy use. For some uses there are no suitable alternatives to electricity—lighting and refrigeration, for example. But clean and affordable alternatives can be found to heating with electricity. Where there is a supply of natural gas, connection subsidies or assistance in financing investment in gas-fired appliances can increase welfare for the poor. Where gas is not available, more efficient and cleaner wood stoves have been successful, as have improvements in building insulation. And in situations where incomes are sufficiently high, temperatures are very cold, and combined heat and power plants are in use, investments in rehabilitating district heating systems can be appropriate.

Raising Tariffs Gradually

If a fiscal crisis does not preclude the practice, raising tariffs gradually can smooth the impact considerably by giving households more time to respond to rising prices, since price elasticity is greater in the long run than in the short. Clearly this is easier for energy-rich Azerbaijan than for energy-poor Armenia, Georgia, or Moldova.

Controlling Consumption

Households must be able to monitor their consumption by use of meters, whether for gas or electricity. Because of technological improvements, more options are now available in metering—for example, smart metering, to allow consumption to be monitored for money spent rather than kilowatt hours consumed—and prepayments. These must be accompanied by financing instruments to give the poor access to systems that will improve their welfare.

Designing and Implementing Successful Reform

Improving Cost Recovery

In addition to mitigating strategies, the studies also shed light on more general factors in the design of the reform that make it successful. Moves to cost recovery will be more palatable and credible to consumers if tariff increases can be more explicitly tied to improvements in service quality. Improving billing and enforcing collections first, before raising tariffs, also increases the likelihood of sustainable reform. Since nonpayment is generally higher among the poor, they will face much higher effective tariff increases than the rich if enforcement and tariffs increase simultaneously.

Outside Factors Affecting Reform

The studies also shed light on institutional and political economy factors that determine the success of reform—factors to some extent beyond the control of the policy makers designing utility sector interventions. A country's macroeconomic condition and resource endowments, the level of competence or corruption in domestic private sector and in government, and the constellation of vested interests that form the political economy backdrop of reform all have significant potential to affect the outcomes of reform—often in unpredictable ways. A resource-poor Armenia, Georgia, or Moldova is pressed to reform far more rapidly than an energy-exporting Azerbaijan. The ebb and flow of political power also plays a key role. With the Shevardnadze government's diminishing credibility in Georgia, corruption and vested interests were given free reign to block reform. In Armenia, too, a year of political instability in 1999 was thought to have compromised efforts to improve cost recovery, while a government far more certain in its position was able to make dramatic inroads in improving collections in 2002.

Designing Suitable Policies

With greater understanding of these factors, policy makers can tailor interventions to mitigate the institutional and political economy risks inherent in the reform environment. In Georgia, the design of the privatization contract allowed the government to distance itself from reform and even blame the private sector for price increases, undermining attempts to improve cost recovery and leaving the private utility to fight political battles. But in a subsequent management contract, an effective mechanism to share risk meant that the Georgian government and utility interests were aligned in encouraging payments. In Armenia, the government took responsibility for improving collections, and thus much of the heat for tariff increases. Another lesson is that though economists and donors can point to an ideal sequence of reforms, reality imposes limitations. Reforms can be unpopular, and small windows of opportunity can often be used around such outside factors as election cycles. Policy makers have to be flexible and adaptive in responding to these external constraints.

Analyzing Reform: The Potential of PSIAs

Generating Better Data and Evidence

The information gained on reform and household responses illustrates the potential for PSIAs to inform policy decisions. In addition to the findings, the studies show the importance of having complete and accurate information as a basis for policy decisions—whether a correct prediction of price responses in response to reform in Armenia, or an accurate estimate of the demand for heat when making large investments in district heating rehabilitation. As these PSIAs were conducted, large gaps were found in the quantitative and qualitative data available—even though monitoring reform performance, and understanding its impact, should be a primary concern of policy makers.

This demonstrates the huge potential of the PSIA methodology for the future of reform design, by highlighting the extent to which reforms are producing outcomes other than those that are commonly assumed or anticipated. In Armenia, the tariff increase was much larger than originally planned because the average monthly payment had not been appropriately calculated. A well-designed program in Georgia had not factored in a pervasive culture of nonpayment and a network where theft was routine and payment unusual. In Moldova, far from hurting the poor, the consumption gap between the poor and the nonpoor was actually narrowing after sector reforms, and the poor in particular appeared to be benefiting from a return to 24-hour service. By bringing empirical evidence to debates characterized by polemics and misperceptions, and by highlighting the need for better data to be collected, such studies have a critical role in designing good policy.

Involving Stakeholders

Since the late 1990s, development institutions have adopted a more participatory approach in their business with client countries. PSIAs are indicative of this change, supporting it in several ways: PSIAs emphasize the distributional impacts of reform; identify the trade-offs between efficiency and equity; account for the concerns of borrowing countries and the constraints facing policy makers; and provide a broad range of stakeholders with the information required for a meaningful policy dialogue. They are thus a critical analytical tool supporting how the World Bank approaches its policy and lending operations. PSIAs further validate this approach by demonstrating how an informed dialogue, based on rigorous empirical evidence, can actually advance reform—by providing clear, empirical answers to the concerns that stakeholders often bring to debates. Such debates have in the past been characterized by ideology and polemics—an obvious example being whether privatization hurts the poor in Moldova.

Building Capacity

The PSIAs go even further in encouraging the participation of developing countries by seeking the involvement of country stakeholders—government counterparts, nongovernmental organizations, utilities, and consumer groups. The analysis is usually conducted in partnership with local consultants who become involved in the PSIA's production. Indeed, the explicit aim is to build capacity to enable countries to conduct such analyses themselves, and to encourage decision making at the local level for reforms whose effects are local. This straightforward approach is making this goal a reality. And it clearly extends beyond the sphere of infrastructure reforms and beyond the ECA region.

There are limitations, of course—the focus on first-order rather than second-order effects and the time frame of the studies, which do not tell the story of the longer term impact of reform. But the ability to analyze the results of reform, ex post and particularly ex ante, is a valuable tool in designing policy.

Lessons for PSIAs

The studies here provide guidelines for using the methodology to analyze infrastructure reforms and undertaking such analysis in the future.

Necessary Steps

The key welfare indicators that such PSIAs must quantify to build a picture of budget shares spent on electricity are household income and expenditure levels and absolute levels of electricity consumption. Additional information is needed on service quality and availability, access to different energy sources, and coping mechanisms. Quantitative data form the empirical backbone of this analysis, but qualitative data are also critical to complete the story provided by the quantitative analysis and to point to the issues that need addressing and the questions that need to be asked.

Adapt to Local Context

The study becomes most valuable when it is adapted to the local political economy and closely tailored to address issues and problems raised by primary stakeholders. Although each of the studies was conceived as a result of a disagreement or impasse in the reform project, conducting qualitative research through focus groups and key informant interviews to inform the quantitative data was a highly effective way of discerning the most pressing concerns of local stakeholders.

Allow Adequate Time and Resources

The studies also demonstrated the importance of allowing adequate time and resources to conduct a careful analysis, since the credibility of the findings rests on the quality of the analysis. To some extent, this finding is in tension with the idea that PSIAs can form part of the cycle of decision making, since the time required for a study may not conform to the Bank or client country's internal framework or budget cycle, and the nature of a crisis may mean that reform is needed in a hurry. But a study conducted without sufficient attention to the quality of the analysis will not provide as useful an input to the debate. Given the controversy surrounding the reforms, and the often contentious nature of the PSIA findings, rigorous analysis becomes all the more important.

Reframe Controversial Issues

Again touching on the political sensitivity of some of the issues that PSIAs address, in Moldova it proved very useful to reframe the issues by asking pointedly neutral questions. Rather than asking whether privatization hurt the poor, the study focused on a straightforward welfare indicator and compared it for different groups. By looking at the issues in a different way, the PSIA can move the debate in a new direction, liberating it from the stalemate induced when it is based on polemics and ideology.

Involve a Broad Range of Stakeholders

The studies also illustrate the value of involving a broad range of stakeholders. Not only is this an important step in understanding the distributional impact of reform—it also makes the PSIA process part of the forum for discussion. In promoting a broad-based dialogue on reform, the study can help clear up misunderstandings, "democratize" the debate, and build consensus on a reform program. Ultimately, this approach can help generate support for a reform process in which more members of society feel ownership.

Ex Post and Ex Ante Approaches

The range of studies in this book—with three ex post studies conducted immediately postreform in Armenia, and several years later in Georgia and Moldova, and an ex ante study in Azerbaijan—conveys an idea of the usefulness of both approaches in informing policy. While ex ante analysis is in many ways preferable when approaching the design of reform, ex post analysis has shown itself extremely useful in keeping a reform program on track when it threatens to derail.

Alternatives to Privatization

Since the hiatus of the 1990s, the enthusiasm for privatization in international financial institutions such as the World Bank has given way to an approach that gives more consideration to public–private approaches.

While this results partly from practical constraints—in a very different world economy where the market appetite of investors has largely evaporated—attitudes have also been tempered by some of the more chastening experiences of privatization. Simply changing ownership has often proved an insufficient, ineffective, or inappropriate tool for turning a failing sector around—as, for example, expecting AES Corporation to transform a culture of nonpayment in Tbilisi in the absence of political support. Utilities that have remained publicly owned have demonstrated that they are capable of becoming efficient and financially sustainable entities, as in Moldova.

The former orthodoxy of privatization has given way to greater consideration of such alternatives as partial privatization or selling off management contracts rather than entire utilities. These alternatives can provide different means to a sustainable sector. This change in stance cannot be traced to any single occurrence or study, but to the body of experience and studies on reform and privatization.

Conclusion

Policy reforms are usually characterized by winners and losers. How to compensate the losers has been the subject of countless debates and studies. This book has taken a close look at the distribution effects of introducing cost recovery to public services that were previously below cost. And it has illustrated how policy options that are widely advocated by economists work in these real-life cases.

Since the late 1990s, Europe and Central Asia (ECA) has seen important—and welcome—changes to the macroeconomic and institutional backdrop for reform. As transition has progressed, poverty levels have declined and living standards have improved across the region. Utility reform, particularly electricity reform, has continued to bring significant improvements to the electricity sector, notwithstanding some of the difficulties met along the way and described in this book. And the climate of reform has changed markedly, with privatization, strongly favored in the mid- to late-1990s, giving way to a management contract approach. Elsewhere, the private sector is returning to parts of ECA. Increases in gas prices must now be considered when recommending gas over electricity as an alternative energy.

Despite these changes, the findings of these studies, undertaken between 1999 and 2004, remain valid for reform today. The timing mismatch between the costs and benefits associated with reform; the fact that the poor consume less energy than the nonpoor, spend a larger percentage of their monthly income on energy, and reduce consumption faster when prices increase; the fact that when the poor have reduced their consumption to basic minimum level, further reductions are unachievable without significant decreases in welfare; and the importance of institutions and political economy aspects in determining reform outcomes, remain critical considerations for policy makers undertaking reform.

Although targeting of social benefit systems has improved markedly since the 1990s, the legacy of systems based on categorical privileges rather than poverty targeting, and the high levels of access to electricity enjoyed by the poor, mean that under the right circumstances tariff-based subsidies can still be more effective than direct transfers in helping the poor access electricity, particularly when a large percentage of the population is poor.

Many studies of utility reforms focus on cases where access to utilities is low and largely concentrated among the nonpoor—a reality in large parts of Africa, Asia, and Latin America. The prescription that flows from this context is a strong argument in favor of direct transfers to poorer consumers, rather than socially regressive tariff-based subsidies. But the findings in this book point to the importance of testing such a prescription against the characteristics of local infrastructure networks. In ECA are found some circumstances that may favor continuation of an alternative system of tariff-based subsidies.

But policy options such as this, or the argument in favor of public interventions to extend access to such efficient alternatives as gas, or the importance for welfare levels of ensuring that service quality improvements accompany tariff increases, are only one part of the lessons learned from this book.

Broadly, the cases looked at here are a powerful testament to the importance of an empirical understanding of the distributional impact on different stakeholders of reform based on quantitative and qualitative analysis. Only by building a comprehensive picture of the behavior of different groups can we understand what reform means to these groups, how and why their behavior changes, and the impact of reform on their welfare levels. The increasing trend to engage in this kind of rigorous analysis to inform the design of successful, sustainable, and politically acceptable reform programs is one more testament to the growing recognition of the importance of this exercise.

Achieving cost recovery remains a pressing need in much of ECA. But hard-won experience, backed by empirical evidence, shows that if welfare losses greatly exceed gains, the social and economic costs of reform can threaten to outweigh the benefits. Moldovans unable to afford more than 55 KWh of electricity per month, who must unplug their refrigerators, minimize use of their television sets, and use low-wattage lightbulbs, or are cut off as a result of being unable to pay their electricity bills, do not perceive the welfare gains from a return to 24-hour service, only the welfare losses from higher prices. In addition to the very real problem for poverty and inequality, this can make reform politically and socially unacceptable. To counter this danger, cost-recovery reforms must be accompanied by measures to facilitate the redistribution of net welfare gains to the most vulnerable members of society to mitigate their welfare losses.

Successful reform depends on many variables in any one country, both within the design of the reform program and beyond it. The reform experience of countries in ECA and elsewhere has spawned much work examining the factors determining success or failure in reform—sequencing, political economy, and institutional factors. It is equally the case that it is easy for a well-conceived reform program to be derailed by factors outside the design of reform. Most often this factor is a lack of political will to support reform, as in Georgia. But improvements in the design of reform aimed at minimizing welfare losses can decrease the potential for organized constituencies to mobilize support against reform. In Bolivia, Georgia, Moldova, and elsewhere, such groups have come perilously close to derailing reform, if not succeeding. Minimizing the likelihood that such constituencies will mobilize is important for successful reform. Putting into action the lessons from the PSIAs in this book, and from those in the future, is central to this effort.

Notes

1. In addition Azerbaijan, Bulgaria, Tajikistan, and Ukraine need to increase tariffs by more than 1.5 cents per KWh. These figures, for 2003, were calculated from World Bank ECA electricity data.

2. World Bank (2004d), p. 1.

ANNEXES

Annex 1. Overview of the Reform Process in Eight ECA Countries

Country	Regulatory development	Corporatization and unbundling of monolithic company	Privatization of distribution	Privatization of generation
Armenia	1997: Energy law established an independent Energy Commission, the Armenian Energy Regulatory Commission.	1997: State-owned enterprise Armenergo unbundled into generation, transmission, and distribution.	2002: Midland Resources Holding (MRH) assumed control of Electricity Distribution Company (EDC) with a management contract. In 2005, MRH sold the company to RAO UES.	2002–03: Ownership of the Hrazdan Thermal Power Plant, the Sevan-Hrazdan Hydro Cascade, and financial control of Medzamor, transferred against US$96 million in state debt forgiveness: Hrazdan TPP transferred to a Russian state company for US$31 million; Sevan Hrazdan Cascade transferred to RAO "Nordic" for US$25 million; and financial management of Medzamor given to another RAO subsidiary, Inter-RAO UES, in exchange for US$40 million in debt for nuclear fuel.
Azerbaijan	Regulatory framework to be established in 2006. Tariff Council has control over tariff policy.	1998: "Azerbaijan Republic Law on electric power engineering" approved. Power grid divided into three parts: State electric energy	2002: Management of four regional distribution companies contracted for a 25-year period to two private companies: Barmek Holding	State-owned enterprise Azernerji manages generation and transmission

(Continued)

Annex 1 (Continued)

Country	Regulatory development	Corporatization and unbundling of monolithic company	Privatization of distribution	Privatization of generation
		enterprise; independent power producers; and power supply enterprises. The state power company, Azerenergy, was turned into a state-owned, closed joint-stock company, with a five-year program for privatization after the company's outstanding debts are paid.	(Turkish) and Baku High Voltage Electrical Equipment.	
Georgia	1997: Electricity Law established an independent regulator, Georgia National Energy Regulatory Commission (GNERC). Georgian Wholesale Electricity Market (GWEM) established in 1999.	1999–2000: State-owned enterprise Sakenergo unbundled.	1998: Tbilisi distribution company Telasi (accounting for 30–50 percent of total national consumption) sold to U.S. company AES. In late 2003, AES sold Telasi to RAO UES of Russia. The other two large distribution companies, UDC and Ajara, are still owned by the state. UDC is under management contract; privatization considered a possibility in the future.	In 2000–02 units, (eight, of which six are not operational) at the thermal generation plant Tbilsresi were sold to AES. AES also managed the Khrami hydrogeneration station. AES sold its assets to RAO UES in 2003. Currently, five generating plants in western Georgia are being prepared for privatization.

Hungary	1993: Policy guidelines created. 1994: Electricity Act; establishment of Hungarian Energy Office (HEO), a regulatory and supervisory body for gas and electricity production, heat production by power stations/large combined heat and power companies; protects consumer interests.	1993–94: MVM Trust unbundled into eight generation companies, one transmission utility, and six distribution companies (EDC).	1995: controlling shares in six EDCs sold to strategic investors (mainly German and French), raising about US$1.1 billion in revenues. 1997–98: remaining shares in EDCs sold through stock market offering.	1995: controlling shares in two generation companies sold to strategic investors; 1996–97: four more generation companies privatized. All power stations have been privatized except the nuclear and an old coal-fired station.
Kazakhstan	1998–99: Law on Natural Monopolies, Law on Electricity, and creation of the regulatory Anti-Monopoly Agency (AMA).	1996: Unbundling of state-owned enterprise Kazakhenergo.	Since 1996, 3 out of 18 distribution companies have been privatized: electricity and heat distribution networks in Almaty region to Tractabel of Belgium in 1996; electricity networks in Karaganda region to National Power of UK in 2000; and networks in the Altai region to AES in 1999.	Since 1996, around 80 to 90 percent of generation assets have been privatized. 1999–2002: Government believed to have sold remaining generation assets to RAO UES.

(Continued)

Annex 1 (Continued)

Country	Regulatory development	Corporatization and unbundling of monolithic company	Privatization of distribution	Privatization of generation
Moldova	1998: Electricity law approved and independent regulatory agency ANRE established.	1997: State energy company Moldenergo unbundled into three generation companies, five distribution, and six other construction and heat companies and a state enterprise responsible for transmission and dispatch.	1999: Sale of three out of five distribution companies (covering more than two-thirds of the market) to Union Fenosa of Spain.	No privatization has yet taken place.
Poland	1997: Energy law laid out reforms and created an independent Energy Regulatory Agency (ERA).	1993: Commercialization and unbundling of PSE (Polish Power Grid Company).	2003: Five distribution companies in Western and Northern Poland consolidated. Future plans include creating three more power distribution enterprises. Privatization is under way in eight power distribution enterprises in Northern and Central Poland.	Consolidation of the generation sector continues with merger of PKE SA and BOT. Plans are underway to merge five other companies that would constitute 26 percent of national installed capacity. No privatization has yet taken place.

Sources: (1) "Private Sector Participation in the Power Sector in ECA Countries: Lessons from the Last Decade." World Bank. 2002. Draft. (2) "Privatization of the Power and Natural Gas Industries in Hungary and Kazakhstan." World Bank. 1999. (3) News sources. (4) Sargsyan G., A. Balabanyan, and D. Hankinson. 2005. "Unexpected Light: Armenia's Experience with Power Sector Reform."

Annex 2. Summary of Household Survey Data[1]

Table A2.1 Power Sector Access, Payment, and Affordability for Urban Households in 2002

Country	Households with access to electricity			Households reported zero electricity expenditures			Electricity expenditures over income		
	Bottom 20%	Top 20%	Total	Bottom 20%	Top 20%	Total	Bottom 20%	Top 20%	Total
Albania	100	100	100	35	7	17	10	5	7
Armenia	98	99	99	52	20	30	10	6	8
Azerbaijan	100	100	100	13	12	12	2	2	2
Belarus	100	100	100	5	3	4	2	1	1
Bulgaria	99	100	100	1	1	1	12	8	10
Georgia	100	100	100	24	10	17	8	4	5
Hungary	100	100	100	2	1	2	7	5	6
Kazakhstan	100	100	100	9	3	4	4	2	2
Kyrgyz Republic	98	99	98	7	2	2	3	2	2
Moldova	95	100	99	30	31	25	9	6	7
Poland	100	100	100	42	28	31	10	5	7
Romania	96	100	99	28	11	15	7	5	6
Russia	100	100	100	19	12	13	2	2	1
Serbia	100	100	100	3	0	1	10	5	7
Tajikistan	100	100	100	22	13	16	3	2	2
Turkey	100	100	100	55	33	43	10	6	8
Ukraine	90	97	96	1	1	1	3	2	2

Table A2.2 Power Sector Access, Payment, and Affordability for Rural Households in 2002

Country	Households with access to electricity			Households reported zero electricity expenditures			Electricity expenditures over income		
	Bottom 20%	Top 20%	Total	Bottom 20%	Top 20%	Total	Bottom 20%	Top 20%	Total
Albania	99	100	100	19	9	13	5	3	4
Armenia	94	99	98	55	42	42	8	4	6
Azerbaijan	99	100	100	16	12	12	2	1	1
Belarus	97	100	99	9	6	6	1	1	1
Bulgaria	94	100	99	1	0	1	8	3	8
Georgia	100	100	100	10	7	8	9	2	4
Hungary	100	100	100	3	1	2	7	5	6
Kazakhstan	100	100	100	4	1	2	3	1	2
Kyrgyz Republic	100	100	100	7	4	6	2	2	2
Moldova	97	100	99	14	11	12	7	4	5
Poland	100	100	100	41	26	32	10	7	8
Romania	86	98	94	34	26	29	6	5	6
Russia	100	100	100	18	10	12	2	1	1
Serbia	99	100	100	4	1	2	7	5	6
Tajikistan	99	99	99	13	7	9	3	1	2
Turkey	100	100	100	50	18	26	10	5	7
Ukraine	94	99	98	0	0	0	3	2	2

Table A2.3 Power Sector Access, Payment, and Affordability for All Households in 2002

Country	Households with access to electricity			Households reported zero electricity expenditures			Electricity expenditures over income		
	Bottom 20%	Top 20%	Total	Bottom 20%	Top 20%	Total	Bottom 20%	Top 20%	Total
Albania	99	100	100	19	9	13	6	4	5
Armenia	94	99	98	55	42	42	10	6	7
Azerbaijan	99	100	100	16	12	12	2	2	2
Belarus	97	100	99	9	6	6	2	1	1
Bulgaria	94	100	99	1	0	1	10	8	9
Georgia	100	100	100	10	7	8	9	3	5
Hungary	100	100	100	3	1	2	7	5	6
Kazakhstan	100	100	100	4	1	2	3	2	2
Kyrgyz Republic	100	100	100	7	4	6	3	2	2
Moldova	97	100	99	14	11	12	8	5	6
Poland	100	100	100	41	26	32	10	6	7
Romania	86	98	94	34	26	29	6	5	6
Russia	100	100	100	18	10	12	2	1	1
Serbia	99	100	100	4	1	2	8	5	6
Tajikistan	99	99	99	13	7	9	3	1	2
Turkey	100	100	100	50	18	26	10	6	7
Ukraine	94	99	98	0	0	0	3	2	2

Table A2.4 Power Sector Affordability Ratio Following Tariff Increase to Full-Cost Recovery

Price elasticity	e = −0.15			e = −0.25			e = −0.35			e = −0.50			e = −1		
Country	Bottom 20%	Top 20%	Total	Bottom 20%	Top 20%	Total	Bottom 20%	Top 20%	Total	Bottom 20%	Top 20%	Total	Bottom 20%	Top 20%	Total
Albania	10	6	8	9	6	8	8	5	7	7	4	5	1	1	1
Armenia	15	9	11	14	8	10	12	7	9	11	6	8	4	3	3
Azerbaijan	4	3	4	2	2	2	0	0	0						
Belarus	3	1	2	3	1	2	2	1	1	1	1	1			
Bulgaria	13	10	12	13	10	11	12	10	11	11	9	10	9	7	8
Georgia	12	4	7	12	4	6	11	4	6	10	3	5	6	2	3
Kazakhstan	6	3	4	5	2	3	4	2	2	2	1	1			
Moldova	10	6	8	10	6	8	9	5	7	9	5	7	7	4	5
Romania	7	6	7	7	6	6	7	6	6	7	6	6	6	5	6
Russia	5	2	3	2	1	1									
Serbia	16	10	12	13	8	10	10	6	8	6	4	4			
Ukraine	6	3	4	4	2	3	3	2	2	1	0	1			

Table A2.5 Gas Sector Access, Payment, and Affordability for Urban Households in 2002

Country	Households with access to network gas			Households reported zero network gas expenditures			Network gas expenditures over income		
	Bottom 20%	Top 20%	Total	Bottom 20%	Top 20%	Total	Bottom 20%	Top 20%	Total
Albania	na	na	na	na	na	na	na	na	na
Armenia	36	45	36	95	82	86	5	5	6
Azerbaijan	92	94	92	21	23	20	2	1	1
Belarus	92	89	90	na	na	na	na	na	na
Bulgaria	1	7	4	0	0	0	3	2	2
Georgia	23	53	37	42	17	26	9	3	4
Hungary	67	84	79	9	7	8	10	5	7
Kazakhstan	30	62	50	51	19	28	3	1	2
Kyrgyz Republic	36	76	60	35	12	17	4	3	3
Moldova	60	79	72	45	29	34	9	3	5
Poland	58	83	74	49	35	38	7	5	5
Romania	55	82	75	35	8	14	6	4	5
Russia	74	65	71	27	18	19	2	0	1
Serbia	5	12	9	32	19	25	7	6	5
Tajikistan	49	51	50	na	na	na	na	na	na
Turkey	0	1	0	—	100	100	—	—	—
Ukraine	74	79	78	14	6	8	4	2	3

n.a. = not available

Table A2.6 Gas Sector Access, Payment, and Affordability for Rural Households in 2002

Country	Households with access to network gas			Households reported zero network gas expenditures			Network gas expenditures over income		
	Bottom 20%	Top 20%	Total	Bottom 20%	Top 20%	Total	Bottom 20%	Top 20%	Total
Albania	na	na	na	na	na	na	na	na	na
Armenia	12	34	22	95	66	67	13	5	5
Azerbaijan	6	11	8	6	6	7	1	2	2
Belarus	95	99	98	na	na	na	na	na	na
Bulgaria	0	1	1	0	0	1	1	1	2
Georgia	8	5	6	40	26	33	7	3	5
Hungary	36	80	58	6	5	7	13	6	8
Kazakhstan	10	9	12	86	57	85	4	0	2
Kyrgyz Republic	10	39	19	100	80	89	—	4	3
Moldova	5	17	10	28	24	24	9	9	8
Poland	13	23	17	41	29	36	7	8	7
Romania	5	16	11	20	14	13	8	9	8
Russia	55	54	56	31	20	23	4	2	3
Serbia	3	8	6	31	13	22	6	5	5
Tajikistan	5	8	7	na	na	na	na	na	na
Turkey	1	27	12	56	22	25	29	7	8
Ukraine	29	48	38	6	2	3	7	5	6

n.a. = not available

Table A2.7 Gas Sector Access, Payment, and Affordability for All Households in 2002

Country	Households with access to network gas			Households reported zero network gas expenditures			Network gas expenditures over income		
	Bottom 20%	Top 20%	Total	Bottom 20%	Top 20%	Total	Bottom 20%	Top 20%	Total
Albania	na	na	na	na	na	na	na	na	na
Armenia	28	40	30	95	76	80	7	5	6
Azerbaijan	57	58	54	18	18	17	2	1	1
Belarus	93	92	92	na	na	na	na	na	na
Bulgaria	1	5	3	0	0	0	3	2	2
Georgia	13	34	22	41	18	27	8	3	4
Hungary	53	83	72	8	6	8	11	5	7
Kazakhstan	15	53	34	67	20	37	3	1	2
Kyrgyz Republic	17	58	33	64	34	43	4	3	3
Moldova	23	49	32	42	28	32	9	4	6
Poland	34	68	51	48	35	38	7	5	5
Romania	21	65	46	32	8	14	7	5	5
Russia	66	63	67	28	18	20	3	1	1
Serbia	4	11	8	32	17	24	7	6	5
Tajikistan	16	24	19	na	na	na	na	na	na
Turkey	1	20	8	56	23	26	29	7	8
Ukraine	57	71	65	13	5	7	4	2	3

n.a. = not available

Table A2.8 District Heating Access, Payment, and Affordability for Urban Households in 2002

Country	Households with access to central heating			Households reported zero central heating expenditures			Central heating expenditures over income		
	Bottom 20%	Top 20%	Total	Bottom 20%	Top 20%	Total	Bottom 20%	Top 20%	Total
Albania	0	1	0	—	0	0	—	22	18
Armenia	4	7	6	100	89	95	—	7	8
Azerbaijan	21	33	24	99	98	99	1	1	1
Belarus	89	94	92	na	na	na	na	na	na
Bulgaria	21	37	31	21	6	10	12	9	11
Georgia	0	0	1	100	92	99	—	12	12
Hungary	18	32	27	10	0	3	15	10	12
Kazakhstan	32	78	60	69	22	32	12	6	8
Kyrgyz Republic	32	72	54	86	27	38	8	4	5
Moldova	60	94	78	99	66	80	19	15	15
Poland	38	67	58	23	10	11	10	7	9
Romania	41	65	57	79	35	45	13	11	11
Russia	85	95	91	36	18	24	4	2	3
Serbia	22	52	40	22	29	27	1	0	1
Tajikistan	10	32	21	91	93	96	4	9	8
Turkey	0	18	5	na	na	na	na	na	na
Ukraine	56	74	64	25	6	12	7	5	6

n.a. = not available

Table A2.9 District Heating Access, Payment, and Affordability for Rural Households in 2002

Country	Households with access to central heating			Households reported zero central heating expenditures			Central heating expenditures over income		
	Bottom 20%	Top 20%	Total	Bottom 20%	Top 20%	Total	Bottom 20%	Top 20%	Total
Albania	0	0	0	—	—	—	—	—	—
Armenia	0	0	0	—	100	100	—	—	—
Azerbaijan	0	0	0	—	—	100	—	—	—
Belarus	50	50	53	na	na	na	na	na	na
Bulgaria	1	4	1	—	20	18	8	10	9
Georgia	0	0	0	53	100	93	8	—	8
Hungary	0	2	0	—	0	0	—	5	5
Kazakhstan	1	4	2	93	56	77	15	6	7
Kyrgyz Republic	1	15	5	0	78	74	5	4	5
Moldova	1	13	5	100	100	100	—	—	—
Poland	3	6	4	19	12	15	9	9	10
Romania	1	2	1	44	36	37	6	7	7
Russia	21	35	27	56	35	44	6	4	4
Serbia	4	22	10	84	83	89	0	2	2
Tajikistan	1	2	1	50	76	77	13	4	5
Turkey	2	45	20	na	na	na	na	na	na
Ukraine	2	3	2	55	3	34	3	3	4

n.a. = not available

Table A2.10 District Heating Access, Payment, and Affordability for All Households in 2002

Country	Households with access to central heating			Households reported zero central heating expenditures			Central heating expenditures over income		
	Bottom 20%	Top 20%	Total	Bottom 20%	Top 20%	Total	Bottom 20%	Top 20%	Total
Albania	0	0	0	—	0	0	—	22	18
Armenia	3	4	3	100	90	95	—	7	8
Azerbaijan	13	19	13	99	98	99	1	1	1
Belarus	79	79	80	na	na	na	na	na	na
Bulgaria	13	29	22	21	6	10	12	9	11
Georgia	0	0	0	70	95	97	8	12	11
Hungary	10	24	18	10	0	3	15	10	12
Kazakhstan	10	65	35	71	22	33	12	6	8
Kyrgyz Republic	9	44	22	81	35	43	7	4	5
Moldova	20	54	32	99	70	82	19	15	15
Poland	20	52	37	22	10	11	10	7	9
Romania	14	49	32	78	35	45	12	11	11
Russia	60	86	74	38	19	26	4	2	3
Serbia	12	43	27	33	38	37	1	0	1
Tajikistan	3	13	6	84	92	94	7	8	7
Turkey	1	38	14	na	na	na	na	na	na
Ukraine	36	56	43	26	6	12	7	5	6

n.a. = not available

Table A2.11 Total Energy Sector (Power, Gas, Heat, Oil, and Wood) Affordability in 2002

Country	Urban			Rural			Total		
	Bottom 20%	Top 20%	Total	Bottom 20%	Top 20%	Total	Bottom 20%	Top 20%	Total
Albania	na	na	na	na	na	na	na	na	na
Armenia	na	na	na	na	na	na	na	na	na
Azerbaijan	4	3	3	5	5	5	4	4	4
Belarus	na	na	na	na	na	na	na	na	na
Bulgaria	18	14	16	12	15	14	16	14	16
Georgia	14	7	10	14	6	9	14	7	9
Hungary	20	13	17	19	14	17	20	14	17
Kazakhstan	10	7	9	7	7	7	9	7	8
Kyrgyz Republic	4	7	6	4	5	4	4	6	5
Moldova	14	14	14	7	10	8	9	12	10
Poland	15	13	14	11	18	13	13	14	14
Romania	na	na	na	na	na	na	na	na	na
Russia	14	7	10	16	9	12	15	8	11
Tajikistan	na	na	na	na	na	na	na	na	na
Turkey	14	13	13	13	12	13	13	12	13
Ukraine	9	7	8	7	7	7	9	7	8

n.a. = not available

Table A2.12 Water Sector Access, Payment, and Affordability for Urban Households in 2002

Country	Households with access to cold water network			Households reported zero cold water expenditures			Cold water expenditures over income		
	Bottom 20%	Top 20%	Total	Bottom 20%	Top 20%	Total	Bottom 20%	Top 20%	Total
Albania	79	96	91	40	14	24	2	1	1
Armenia	94	99	97	97	88	92	2	2	3
Azerbaijan	89	87	86	21	27	23	1	1	1
Belarus	87	93	91	na	na	na	na	na	na
Bulgaria	93	100	99	15	8	8	5	2	3
Georgia	90	97	94	70	45	56	2	1	1
Hungary	94	99	98	23	21	20	5	3	4
Kazakhstan	68	93	86	29	10	14	2	1	1
Kyrgyz Republic	63	91	80	44	14	21	1	1	1
Moldova	61	96	78	69	45	52	4	2	3
Poland	97	100	99	37	18	21	5	2	3
Romania	77	97	92	42	9	18	6	4	5
Russia	94	98	96	32	15	20	2	1	1
Serbia	96	100	99	na	na	na	na	na	na
Tajikistan	82	87	82	34	28	31	3	2	2
Turkey	71	96	87	74	57	62	5	2	3
Ukraine	84	95	89	19	6	9	2	1	2

n.a. = not available

Table A2.13 Water Sector Access, Payment, and Affordability for Rural Households in 2002

Country	Households with access to cold water network			Households reported zero cold water expenditures			Cold water expenditures over income		
	Bottom 20%	Top 20%	Total	Bottom 20%	Top 20%	Total	Bottom 20%	Top 20%	Total
Albania	33	53	39	17	21	24	2	1	1
Armenia	87	88	86	97	95	95	3	1	2
Azerbaijan	27	32	32	27	9	13	1	1	1
Belarus	53	52	55	na	na	na	na	na	na
Bulgaria	86	99	93	12	7	7	5	3	4
Georgia	66	79	74	78	83	80	3	0	1
Hungary	85	99	94	21	18	21	4	3	4
Kazakhstan	34	37	33	3	13	6	2	1	1
Kyrgyz Republic	10	47	22	37	21	23	0	1	1
Moldova	2	8	4	52	33	38	4	2	3
Poland	90	97	94	65	53	58	4	3	3
Romania	7	23	14	42	26	30	4	3	3
Russia	57	73	65	52	41	45	2	1	1
Serbia	71	92	83	na	na	na	na	na	na
Tajikistan	21	36	29	16	18	17	2	1	2
Turkey	98	100	99	50	26	33	6	3	4
Ukraine	35	35	34	29	23	27	2	1	1

n.a. = not available

Table A2.14 Water Sector Access, Payment, and Affordability for All Households in 2002

Country	Households with access to cold water network			Households reported zero cold water expenditures			Cold water expenditures over income		
	Bottom 20%	Top 20%	Total	Bottom 20%	Top 20%	Total	Bottom 20%	Top 20%	Total
Albania	48	76	61	29	16	24	2	1	1
Armenia	91	94	92	97	91	93	2	2	2
Azerbaijan	64	63	62	22	23	21	1	1	1
Belarus	78	79	80	na	na	na	na	na	na
Bulgaria	90	100	97	14	8	8	5	3	4
Georgia	75	90	84	75	58	66	2	1	1
Hungary	90	99	97	22	21	21	5	3	4
Kazakhstan	44	84	63	15	10	12	2	1	1
Kyrgyz Republic	24	69	42	42	16	22	1	1	1
Moldova	21	52	32	67	44	51	4	2	3
Poland	93	99	97	51	26	35	4	3	3
Romania	30	79	56	42	10	19	6	4	5
Russia	79	94	88	37	18	25	2	1	1
Serbia	82	98	92	na	na	na	na	na	na
Tajikistan	36	54	43	27	23	24	3	1	2
Turkey	85	99	94	59	34	44	5	2	4
Ukraine	66	80	71	21	8	12	2	1	2

n.a. = not available

Note

1. All data reported in this annex are derived by the authors from 2002 household budget data.

Annex 3. Converting Energy Prices into Cost per Effective Btu

Table A3.1 Calculation of Cost per Effective Btu

Fuel	Original	Household price in Tbilisi, December 2002(a)	Energy content (Btu per original unit)(b)	Cost per mmBtu (GEL)	Efficiency (household use)(c)	Cost per effective mmBtu(GEL)	Dollars per effective mmBtu(d) (US$)
[1]	[2]	[3]	[4]	[5]=10⁻⁶[3]/[4]	[6]	[7]=[5]*[6]	[8]
Natural Gas	m³	0.270	3,412	7.65	70%	10.93	5.08
Electricity	KWh	0.137	35,300	40.15	90%	44.61	20.75
Kerosene	liter	0.790	32,934	24.04	40%	60.09	27.95
LPG	kg	1.400	42,854	32.67	70%	46.67	21.71
Fuel wood	m³	22.563	7,165,200	3.15	20%	15.74	7.32

Source: Authors' calculations.

a. Energy prices (except wood) from State Department of Statistics. Price of wood from USAID/Save the Children.

b. World Bank staff estimates.

c. World Bank staff estimates.

d. Exchange rate was 2.15 in December 2002.

Note: LPG is liquefied petroleum gas.

Annex 4. Combined Household Survey and Utility Data for Four Countries

Table A4.1 Summary of Combined Household Survey and Utility Data for Four Countries

	Average aggregate collection rate paid/billed (%)	Price of electricity per KWh UScents/KWh	Monthly KWh (utilities records) KWh	Monthly electricity expenditures (stated in HBS) US$/month	Monthly income US$/month[a]	Electricity expenditures as percent of income Stated (%)[b]
Georgia (Tbilisi)[c]						
2000, q1	22	4.55	205	2.7	168	2.0
q2	24	4.55	207	2.9	138	2.5
q3	31	4.68	179	2.7	171	2.3
q4	35	4.95	146	3.5	171	2.9
2001, q1	62	4.73	146	4.2	182	2.7
q2	56	4.73	156	4.9	169	3.5
q3	64	4.73	128	5.2	164	4.1
q4	73	5.15	143	5.6	169	4.0
2002, q1	133	5.64	173	5.8	165	4.2
q2	77	5.64	170	6.0	164	4.4
q3	73	5.64	139	4.0	172	5.9
q4	75	5.64	151	5.9	189	4.4
Moldova (Union Fenosa service area)						
2001, q1	100	5.05	43	3.0	97	3.4
2	100	5.05	37	2.1	47	4.1
3	99	5.05	41	2.9	53	5.7
4	100	5.05	41	2.5	84	4.7
5	100	5.05	39	2.2	73	2.5
6	100	5.05	29	2.0	48	4.5

(Continued)

Table A4.1. (Continued)

	Average aggregate collection rate paid/billed (%)	Price of electricity per KWh UScents/KWh	Monthly KWh (utilities records) KWh	Monthly electricity expenditures (stated in HBS) US$/month	Monthly income US$/month[a]	Electricity expenditures as percent of income Stated (%)[b]
7	100	5.05	29	2.2	50	5.5
8	99	5.04	47	3.2	73	6.0
9	100	5.01	51	3.0	72	5.4
10	100	5.03	56	3.0	90	4.5
11	100	5.25	60	3.2	72	5.1
12	100	5.26	57	3.0	72	4.9
2002, q1	100	4.99	62	3.2	76	4.3
2	100	4.99	53	2.8	73	4.9
3	99	5.00	50	2.6	70	4.7
4	100	5.00	49	2.7	69	5.0
5	100	4.98	53	2.6	74	4.0
6	98	4.99	45	2.5	73	4.4
7	99	4.98	50	2.6	77	4.3
8	100	4.99	50	2.8	87	4.1
9	99	5.26	52	3.0	109	3.6
10	100	5.25	52	2.6	99	3.4
11	100	5.26	60	3.1	94	3.8
12	100	5.27	64	3.5	102	3.9
2003, q1	98	5.10	69	3.0	98	3.6
2	100	5.13	58	4.0	94	4.5
3	99	5.13	65	3.3	90	3.9
4	99	5.15	56	3.0	100	3.5
5	100	5.13	56	3.3	95	3.8

6	99	5.13	55	3.1	92	4.1
7	96	5.17	52	3.0	104	3.7
8	97	5.61	51	2.7	102	3.4
9	96	5.55	51	3.4	127	3.9
10	95	5.54	60	2.9	105	3.4
11	85	5.51	61	3.3	91	4.5
Armenia (Yerevan) 3/						
June–Dec. 98	89	3.80	173	5.9	100	9.0
Azerbaijan 2002 (all months) Baku, only metered households						
Poorest						
20%	65	1.96	190	2.2	123	2.1
2	61	1.96	202	2.1	137	1.9
3	74	1.96	192	2.3	154	1.9
4	68	1.96	201	2.4	161	1.9
Richest						
20%	81	1.96	200	2.6	189	2.2
Total	71	1.96	198	2.3	158	2.0

Source: Calculated from household survey data and utility company billing records.

a. Income proxied by total monthly household expenditures.

b. In Armenia and in Azerbaijan, electricity expenditures shown here are not stated in the survey, but calculated as an average monthly electricity payment from the utility company records.

c. Decreasing electricity consumption despite increasing income may be due to rationing.

Annex 5. Changes in Generation Mix in the Past Decade and Price and Income Elasticity of Demand Estimates

Figure A5.1 Changes in Generation Mix in the Past Decade

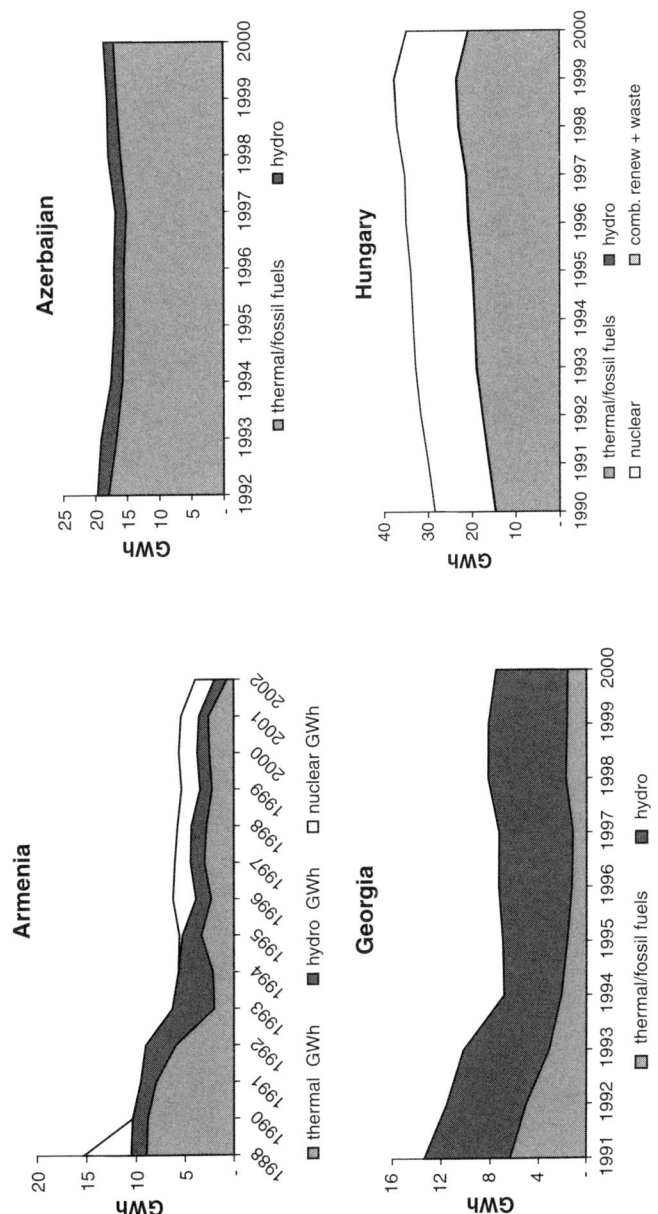

(Continued)

207

Figure A5.1 (Continued)

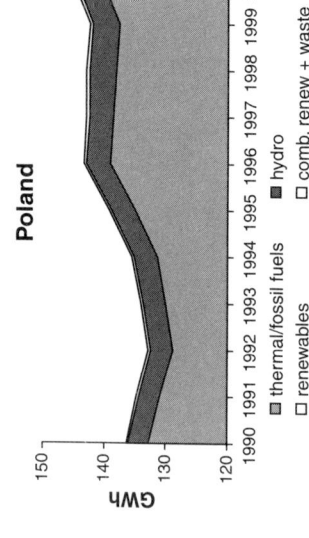

Poland

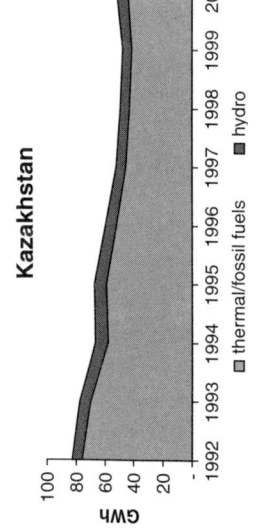

Kazakhstan

Source: Lampietti 2004.

Table A5.1. Empirical Estimates of Price and Income Elasticity of Residential Electricity Demand in Developing Countries

Country	Price elasticity	Income elasticity	Source
Ethiopia	−0.74	1.005	Kebede, Bereket, Almaz Bekele, and Elias Kedir. 2002. "Can the Urban Poor Afford Modern Energy? The Case of Ethiopia." *Energy Policy* 30.
Greece	−0.41	1.56	Hondroyiannis, George. 2004. "Estimating Residential Demand for Electricity in Greece." *Energy Economics*.
India	−0.42 (winter) −0.51 (monsoon −0.29 (summer)	0.6–0.64	Filippini, Massimo, and Shonali Pachauria. 2004. Elasticities of electricity demand in urban Indian households. *Energy Policy* 32: 429–436.
Norway	−0.5 (short-run)	0.2	Nesbakken, Runa. 1999. "Price sensitivity of residential energy consumption in Norway." *Energy Economics* 21.
Taiwan	−0.15	1.04	Holtedahl, Pernille, and Frederick L. Joutz. 2004. "Residential Electricity Demand in Taiwan." *Energy Economics* 26.
United Kingdom	−0.5	0.5	Manning, D. N. 1988. "Household demand for energy in the UK." *Energy Economics* January.
United States	−0.5	0.62	Silk, Julian I., and Frederick L. Joutz. 1997. "Short and long-run elasticities in U.S. residential electricity demand: A co-integration approach." *Energy Economics* 19.
United States	−0.27		Wills, John. 1981. "Residential demand for electricity." *Energy Economics* October.

References

Alam, A., M. Murthi, R. Yemtsov, and others. 2005. *Growth, Poverty, and Inequality—Eastern Europe and the Former Soviet Union.* Washington, DC: World Bank.

ANRE (National Energy Regulatory Agency). 2002. *Report on National Agency for Regulation of Energy Activity During 2001.* Chisinau.

ANRE. 2003. *Report on National Agency for Regulation of Energy Activity During 2002.* Chisinau.

BBC News Online. 2006. "Russia Blamed for 'Gas Sabotage.'" January 22, 2006.

Besant-Jones, John E. 2006. *Lessons and Sourcebook for Reforming Power Markets in Developing Countries.* Washington, DC: World Bank.

Bhatnagar, B. 2001. "Filipino Report Card on Pro-poor Services." Report No. 22181-PH, Environment and Social Development Sector Unit, East Asia and Pacific. World Bank, Washington, DC.

Birdsall, N., and J. Nellis, eds. 2005. *Reality Check: The Distributional Impact of Privatization in Developing Countries.* Center for Global Development, Washington, DC.

Birdsall, N., and J. Nellis. 2003. "Winners and Losers: Assessing the Distributional Impact of Privatization." *World Development* 31(10): 1617–33.

Bourguignon, F., and L. A. Pereira da Silva. 2003. "The Impact of Economic Policies on Poverty and Income Distribution: Evaluation Techniques and Tools." Washington, DC: World Bank. www.worldbank.org/PSIA.

Bucknall, J. 1999. "Poland—Program for Air Protection in Silesia Project." Project Information Document, World Bank, Washington, DC.

Bucknall, J., and G. Hughes. 2000. "Poland—Complying with EU Environmental Legislation." Technical Paper No. 454, World Bank, Washington, DC.

Castel, Paulette. Forthcoming. "Moldova: Public and Private Safety Net." In *Recession, Recovery, and Poverty in Moldova*. Vol. 2. Report No. 28024-MD. ECSPE, World Bank, Washington, DC.

Chisari, O., A. Estache, and C. Romero. 1999. *Winners and Losers from Utility Privatization in Argentina: Lessons from a General Equilibrium Model*. Policy Research Working Paper No. 1824, November 30, 1999, World Bank, Washington, DC.

CISR (Counterpart International/Center for Strategic Studies and Reforms). 2002. "Evaluation of Social Assistance to Population through the Nominative Targeted Compensations." Low-Income Energy and Social Assistance Project (LIESAP). Chisinau.

Clarke, G. R. G., and S. J. Wallsten. 2002. *Universal(ly Bad) Service: Providing Infrastructure Services to Rural and Poor Urban Consumers*. Policy Research Working Paper No. 2868, July 19, 2002, World Bank, Washington, DC.

Cornillie, J., and S. Frankhauser. 2002. "The Energy Intensity of Transition Countries." Working Paper 72, European Bank for Reconstruction and Development.

Coudouel, Aline, and Stefano Paternostro, eds. 2005. *Analyzing the Distributional Impact of Reforms: A Practitioner's Guide to Trade, Monetary, and Exchange Rate Policy, Utility Provision, Agricultural Markets, Land Policy and Education. Volume 1*. Washington, DC: World Bank.

COWI A/S. 2002a. *Urban Heating Strategy Development for Republic of Armenia*. Draft Final Consultant Report to the Government of Armenia.

COWI A/S. 2002b. *Development of Heat Strategies for the Kyrgyz Republic*. Draft Final Consultant Report.

Dalmazzo, A., and G. de Blasio. 2003. "Resources and Incentives to Reform." Staff Papers 50(2), International Monetary Fund, Washington, DC.

Department of Privatization of the Republic of Moldova. http://www.privatization. md/privatization/privatization_companies/strategic_projects/wine_tobac-co_energy/energy/.

Department of Statistics, Yerevan. 1999. "Socio-economic Situation of the Republic of Armenia," January–December. Monthly Statistical Analytical Report 2000. Government of Armenia.

Dodonu, Veaceslav. 1999. "Moldova's Power Sector: Challenges and Opportunities." www.bisnis.doc.gov/bisnis/bisdoc/000113power.htm.

EBRD (European Bank for Reconstruction and Development). 2001 "Transition Report 2001: Energy in Transition." London.

Economist Intelligence Unit. 2004. "Moldova Country Report. February 2004." www.eiu.com.

Environmental Resources Management. 2002. *Urban Heating Strategy for Armenia: Demand Analysis*. Draft Report. London.

Esanov, A., M. Raiser, and W. Buiter. 2001. "Nature's Blessing or Nature's Curse: The Political Economy of Transition in Resource Based Economies." Working Paper 65, European Bank of Reconstruction and Development. London.

ESMAP (Energy Sector Management Assistance Program). 2000. *Increasing the Efficiency of Heating Systems in Central and Eastern Europe and the Former Soviet Union*. Report 234/00, ESMAP, World Bank, Washington, DC.

ESMAP. 2001. *Improved Space Heating Stoves for Ulan Baatar (Mongolia)*. Implementation Completion Report, ESMAP, World Bank, Washington, DC.

Estache, A., V. Foster, and Q. Wodon. 2001. *Accounting for Poverty in Infrastructure Reform*. World Bank Institute, World Bank, Washington, DC.

Flaig, Gebhard. 1990. "Household Production and the Short- and Long-Run Demand for Electricity." *Energy Economics* April 1990.

Foreign Investment Advisory Service Report on Azerbaijan. 2002. International Finance Corporation, Washington, DC.

Foster V., E. Tiongson, and C. R. Laderichi. 2005. In *Analyzing the Distributional Impact of Reforms: A Practitioner's Guide to Trade, Monetary, and Exchange Rate Policy, Utility Provision, Agricultural Markets, Land Policy, and Education. Volume 1*. A. Coudouel and S. Paternostro, eds. World Bank, Washington, DC.

Freinkman, L., G. Gyulumyan, and A. Kyurumyan. 2003. "Quasi-Fiscal Activities, Hidden Government Subsidies, and Fiscal Adjustment in Armenia." World Bank, Washington, DC.

Freund, C. L., and C. I. Wallich. 1995. "Raising Household Energy Prices in Poland: Who Gains? Who Loses?" Policy Research Working Paper 1495, World Bank, Washington, DC.

Freund, C., and C. Wallich. 1996. "The Welfare Effect of Raising Household Energy Prices in Poland." *The Energy Journal* 17:1.

Galal, A., L. Jones, P. Tandon, and I. Vogelsang. 1994. *Welfare Consequences of Selling Public Enterprises: An empirical analysis*. World Bank: Oxford University Press.

Guasch, J. L. 2004. "Granting and Renegotiating Infrastructure Concessions—Avoiding the Pitfalls." World Bank, Washington, DC.

Hellman, J. S. 1998. "Winners Take All: The Politics of Partial Reform in Post-Communist Transitions." *World Politics* 50(2).

Hill, Fiona, and Clifford G. Gaddy. 2003. *The Siberian Curse: How Communist Planners Left Russia Out in the Cold.* Washington, DC: Brookings Institution Press.

IMF (International Monetary Fund). 2001a. "Republic of Armenia: Recent Economic Developments and Selected Issues." Country Report 01/78. IMF, Washington, DC.

IMF. 2001b. "Republic of Moldova: Recent Economic Developments and Selected Issues." Country Report No. 01/22. IMF, Washington, DC.

IMF. 2001c. "Republic of Georgia: Recent Economic Developments and Selected Issues." Country Report 01/211. IMF, Washington, DC.

Institute for Public Policy/CIVIS. 2004. "Barometer of Public Opinion April–May 2004." Public Opinion Poll Press Release. Chisinau, Moldova.

International Energy Agency. 2000. "Energy Balances of OECD and Non-OECD Countries." www.iea.org.

———. Various years. "Energy Balances of OECD and Non-OECD Countries." www.iea.org.

IPA Energy Consulting. 2003. "South East Europe Power Sector Affordability." November. Edinburgh, Scotland.

Junge, Nils, and Julian A. Lampietti. 2006. In *Poverty and Social Impact Analysis of Reforms: Lessons and Examples from Implementation,* ed. A. Coudouel, A. A. Dani, and S. Paternostro. Washington, DC: World Bank.

Kaiser, M. J. 1999. "Energy Utilization Patterns in the Republic of Armenia I. Residential Sector." *Energy Sources* 21 (9).

Katyshev, S., and G. Mandrovskaya. 2003. "Social and Environmental Impact of Electricity Reform in Kazakhstan." ECSSD, World Bank, Washington, DC.

Kennedy, D. 1996. "Competition in the Power Sector of Transition Economies." Working Paper 41, European Bank of Reconstruction and Development, London.

Kennedy, D. 2003. "Power Sector Regulatory Reform in Transition Economies: Progress and Lessons Learned." Working Paper 78, European Bank of Reconstruction and Development, London.

Komives, K., D. Whittington, and X. Wu. 2001. "Infrastructure coverage and the poor: A global perspective." Policy Research Working Paper 2551, World Bank, Washington, DC.

Komives, K. D., Vivien Foster, Jonathan Halpern, and Quentin Wodon. 2005. *Water, Electricity, and the Poor: Who Benefits from Utility Subsidies?* Washington, DC: World Bank.

Krishnaswamy, V., and G. Stuggins. 2003. "Private Participation in the Power Sector in Europe and Central Asia, Lessons from the Last Decade." Working Paper No. 8, World Bank, Washington, DC.

Lampietti, J. A., ed. 2004. *Power's Promise: Electricity Reforms in Eastern Europe and Central Asia*. Working Paper No. 40, World Bank, Washington, DC.

Lampietti, J. A., and A. Meyer. 2002. "Coping with the Cold: Heating Strategies for Europe and Central Asia's Urban Poor." Technical Paper No. 529, World Bank, Washington, DC.

Lampietti, J. A., A. A. Kolb, S. Gulyani, and V. Avenesyan. 2001. "Utility Pricing and the Poor: Lessons from Armenia." Technical Paper No. 497, World Bank, Washington, DC.

Lampietti, J. A., and B. Kropp. 2002. "Climbing Down the Energy Ladder? Household Energy Trends in Eastern Europe and Central Asia." ECSSD, World Bank, Washington, DC.

Lampietti, J. A., H. Gonzalez, E. Hamilton, M. Wilson, and S. Vashakmadze. 2003. "Revisiting Reform: Lessons from Georgia." ECSSD, World Bank, Washington, DC.

Lengyel, L. 2003. "Status of Power Sector in Hungary." Eurocorp Commerz Ltd., ECSSD, World Bank, Washington, DC.

Linn, Johannes, and Roy With. Bahl. 1992. *Urban Public Finance in Developing Countries*. New York: Oxford University Press.

Lovei, L., E. Gurenko, M. Haney, P. O. Keefe, and M. Shkaratan. 2000. *Scorecard for Subsidies: How Utility Subsidies Perform in Transition Economies*. Washington, DC: World Bank.

Lvovsky, K. 2000. "Environmental Costs of Fossil Fuels." Environment Department Papers No. 78, Pollution Management Series, World Bank, Washington, DC.

Markandya, A., M. Jayawardena, and R. Sharma. 2001. "The Impact of Infrastructure Investments on Measured Poverty." A Viewpoint Note, unpublished manuscript, World Bank, Washington, DC.

McKenzie, D., and D. Mookherjee. 2002. "Distributive Impact of Privatization in Latin America: An Overview of Evidence from Four Countries." *Economia* Spring 2003: 161–218.

Moldovan Economic Trends. 2003. "Economic Trends Quarterly Issue." July–September. www.met.dnt.md.

Nadiradze, N. 2003. "Study of Social and Environmental Impacts of Electricity Reform in Georgia." ECSSD, World Bank, Washington, DC.

Paul, S. 1994. "Does Voice Matter? For Public Accountability, Yes." Policy Research Working Paper 1388, World Bank, Washington, DC.

Paul, S. 1998. "Making Voice Work: The Report Card on Bangalore's Public Service." Policy Research Working Paper 1921, World Bank, Washington, DC.

Petri, M., G. Taube, and A. Tsyvinski. 2002. "Energy Sector Quasi-Fiscal Activities in the Countries of the Former Soviet Union." IMF Working Paper WP/02/60, IMF, Washington, DC.

Public Expenditure Database for the Transition Countries. World Bank (2002).

Republic of Moldova. 2002. *Annual Social Report 2001*. Ministry of Labor and Social Protection, Chisinau.

Republic of Moldova. 2003. *Annual Social Report 2002*. Ministry of Labor and Social Protection, Chisinau.

Rio+10. 2001. "Report on the Rio+10." National Assessment Workshop Moldova, Chisinau, June 20, 2001.

Saavalainen, T. O., and J. ten Berge. 2003. "Energy Conditionality in Poor CIS Countries." IMF, Washington, DC.

Sachs, J., and A. Warner. 1995. "Natural Resource Abundance and Economic Growth." NBER Working Paper No. 5398, National Bureau of Economic Research, Cambridge, Mass.

Sargsyan, Gevorg, Ani Balabanyan, and Denzel Hankinson. 2005. "Unexpected Light: Armenia's Experience with Power Sector Reform." World Bank, Washington, DC.

State Department for Statistics of Georgia. 2001. "Poverty Monitoring in Georgia: Annual Report 2000." Tbilisi.

STC (Save the Children). 2002. "Multi-Sectoral National Survey of Households in Georgia."

Subramanian, S., and A. Deaton. 1996. The Demand for Food and Calories. *Journal of Political Economy* 104:1.

SwedPower/FVB. 2001. *Strategic Heating Options for Moldova*. Final Consultant Report.

The Economist. 2002. "Russian Electricity: In Need of Shock Therapy." August 31, pp. 50–51.

UNECE (United Nations Economic Commission for Europe) Co/FAO (Food and Agriculture Organization). 2000. "Forest Resources of Europe, CIS, North America, Australia, Japan and New Zealand: Contribution to the Global Forest Resources Assessment 2000." Geneva Timber and Forest Study Papers 17. United Nations, New York and Geneva.

United Nations Environment Programme. 2002. "Caucasus Environment Outlook." UNEP, Nairobi.

U.S. Department of Energy, Energy Information Administration. 1997. *Annual Energy Consumption Percentiles, 1997*. www.eia.doe.gov.

Valiyev, V. 2003. "State of Azerbaijan Republic Electric Power Industry (1990–2002)." Shems Energy Ltd., Baku.

Vashakmadze, E., and V. Kvekvetsia. 2000. "Georgian Power Sector Deficit." Sector Note, ECSSD, World Bank, Washington, DC.

Wodon, Q., M. I. Ajwad, J. Baker, R. Jayasuriya, C. Siaens, and J. P. Tre. 2003. "Poverty and Public Spending in Latin America." World Bank, Washington, DC.

World Bank. 1993a. "Armenia: Energy Sector Review." Washington, DC.

World Bank. 1993b. "Energy Efficiency and Conservation in the Developing World: The World Bank's Role." World Bank Policy Paper, Washington, DC.

World Bank. 1996a. "Improving Social Assistance in Armenia." Washington, DC.

World Bank. 1996b. "Republic of Moldova: Energy Project." Staff Appraisal Report, Washington, DC.

World Bank. 1997a. "Georgia: Power Rehabilitation Project." Staff Appraisal Report, Washington, DC.

World Bank. 1997b. "Poland—Country Economic Memorandum: Reform and Growth on the Road to the EU." World Bank Economic Report, July 1997, Washington, DC.

World Bank. 1998. "Energy in Europe and Central Asia: A Sector Strategy for the World Bank Group." Discussion Paper No. 393, Washington, DC.

World Bank. 1999a. "Privatization of Power and Natural Gas Industries in Hungary and Kazakhstan." Technical Paper No. 451, Washington, DC.

World Bank. 1999b. "Georgia Poverty and Income Distribution." Report No. 19348-GE, Washington, DC.

World Bank. 1999c. "Hungary, on the Road to the European Union." World Bank Country Study, November 1999, Washington, DC.

World Bank. 1999d. "Proposed Energy Sector Adjustment Credit (Georgia)." Washington, DC.

World Bank. 2000a. "Kazakhstan Public Expenditure Review—I, II, and III." Report No. 20489-KZ, Poverty Reduction and Economic Management Unit, Europe and Central Asia, Washington, DC.

World Bank. 2000b. "Making Transition Work for Everyone: Poverty and Inequality in Europe and Central Asia." Washington, DC.

World Bank. 2000c. *World Development Indicators*. Washington, DC.

World Bank. 2000d. "Estonia: Implementation Completion Report for a District Heating Rehabilitation Project." Report No. 20631, Washington, DC.

World Bank. 2000e. "Maintaining Utility Services for the Poor: Policies and Practices in Central and Eastern Europe and the Former Soviet Union." Washington, DC.

World Bank. 2001. *World Development Report 2000/2001: Attacking Poverty*. New York: Oxford University Press.

World Bank. 2002a. "Description of the Existing Power Networks in Armenia— Bank Mission Aide Memoirs." ECSIE, Washington, DC.

World Bank. 2002b. "Implementation Completion Report. Republic of Moldova Energy Project." Report No. 24446-MD, ECSIE, Washington, DC.

World Bank. 2002c. "Georgia Public Expenditure Review." Report No. 22913-GE, Poverty Reduction and Economic Management Unit, Europe and Central Asia, Washington, DC.

World Bank. 2002d. "Private Sector Development in the Electric Power Sector: A Joint OED/OEG/OEU Review of the World Bank's Assistance in the 1990s." Washington, DC.

World Bank. 2002e. "Georgia Poverty Update." Report No. 22350-GE, Washington, DC.

World Bank. 2003a. *World Development Indicators*. CD-ROM, Washington, DC.

World Bank. 2003b. "Armenia Public Expenditure Review." Report No. 24434-AM, Poverty Reduction and Economic Management Unit, Europe and Central Asia, Washington, DC.

World Bank. 2003c. "Azerbaijan Public Expenditure Review." Report No. 25233-AZ, Poverty Reduction and Economic Management Unit, Europe and Central Asia, Washington, DC.

World Bank. 2003d. "Moldova Public Economic Management Review." Report No. 25423-MD, Poverty Reduction and Economic Management Unit, Europe and Central Asia, Washington, DC.

World Bank. 2003e."Project Appraisal Document. Moldova Energy II." Washington, DC.

World Bank. 2003f. "Azerbaijan Rural Investment Project Social Assessment." Washington, DC.

World Bank. 2003g, "A User's Guide to Poverty and Social Impact Analysis." Washington, DC. www.worldbank.org/psia.

World Bank. 2004a. "Public and Private Sector Roles in the Supply of Electricity Services." Washington, DC.

World Bank. 2004b. "World Development Report—Making Services Work for Poor People." Washington, DC.

World Bank. 2004c. *Recession, Recovery and Poverty in Moldova*. Vols. I and II. Report No. 28024-MD. ECSPE, Washington, DC.

World Bank. 2004d. "Good Practice Note: Using Poverty and Social Impact Analysis to Support Development Policy Operations." June 21, Washington, DC.

World Bank. 2004e. "Armenia—Country Assistance Strategy." June 30, Washington, DC.

World Bank. 2004f. "Moldova Sharing Power: Lessons Learned from the Reform and Privatization of Moldova's Electricity Sector." December 10, Washington, DC.

World Bank. 2004g. "Azerbaijan Raising Rates: Short-Term Implications of Residential Electricity Tariff Rebating." December 10, Washington, DC.

World Bank. 2005a. "Country Partnership Strategy for Georgia." August 18, Washington, DC.

World Bank. 2005b. *World Development Indicators.* Washington, DC.

World Bank. 2006. *World Development Report 2006: Equity and Development.* Washington, DC.

WHO (World Health Organization). 2002. "Reducing Risk, Promoting Healthy Life." World Health Report, Geneva.

Index

traditional fuel usage. *See* fuel substitutes-
transparency and accountability,
8, 153
Turkmenistan, 36

U

Ukraine, 90, 104, 161*n*
Union Fenosa, 13, 91, 92*b*, 98–101
United Energy Distribution Company
(UEDC), 82
United Kingdom, 4, 112
United Nations Environment Program
(UNEP), 157
United States, 93–94, 112
urban households
district heating access, payment, and
affordability by country, 192*t*
electricity access, payment, and
affordability by country, 185*t*
gas sector access, payment, and
affordability by country, 189*t*
poor. *See* heating strategies for urban
poor
total energy sector affordability by
country, 195*t*
water sector access, payment, and
affordability, 196*t*
U.S. Agency for International
Development (USAID), 75, 86*n*
Uzbekistan, 166, 166*f*

V

value–added tax (VAT), 82
volume–differentiated tariff, 151
voucher privatization, 14

W

Washington Consensus, 14
water, heating of. *See* hot water
water sector access, payment, and affordability
all households, 198*t*
rural households, 197*t*
urban households, 196*t*
welfare impact of reforms, 20, 148–52
See also poverty and social impact
analyses (PSIAs); *specific countries*
Winter Heat Assistance Program (WHAP,
Georgia), 75, 77–78, 80
wood. *See* fuel substitutes
World Bank
Armenia and, 45, 60*n*, 61*n*
Georgia and, 64
heating for urban poor and, 125, 126
privatization and, 174–75
on PSIAs, 30, 167
role of, 4
transparency and accountability and, 153
Ukraine and, 161*n*
*World Development Report 2004: Making
Services Work for Poor People* (World
Bank), 153

Eco-Audit

Environmental Benefits Statement

The World Bank is committed to preserving endangered forests and natural resources. The Office of the Publisher has chosen to print *People and Power* on 30% post-consumer chlorine-free recycled fiber paper in accordance with the recommended standards for paper usage set by the Green Press Initiative—a nonprofit program supporting publishers in using fiber that is not sourced from endangered forests. By using this paper, the following were saved: 7 trees, 341 pounds of solid waste, 2,658 gallons of water, 640 pounds of greenhouse gases, and 5 million BTUs of total energy. For more information, visit www.greenpressinitiative.org.